Irrigation and Agricultural Drainage Engineering

Irrigation and Agricultural Drainage Engineering

Editor

Aleix Campo

scitus
academics

Irrigation and Agricultural Drainage Engineering
Edited by **Aleix Campo**

ISBN: 978-1-68117-541-6
Library of Congress Control Number: 2016942941

SCITUS Academics LLC,
www.scitusacademics.com
Box No. 4766, 616 Corporate Way,
Suite 2, Valley Cottage,
NY 10989

Printed in United States of America on Acid Free Paper

Preface

Drainage Water Management is a new practice in which water control structures are installed in the main drain lines to hold water back and allow farmers to drain only as needed.Irrigation Engineering is important since it helps determine future Irrigation expectations. Irrigation has been a central feature of agriculture for over 5000 years, and was the basis of the economy and society of numerous societies, ranging from Asia to Arizona. Irrigation can be termed as the artificial process of applying water to the soil to help in growing agricultural crops or maintaining the landscapes when there is shortage of natural water by rain. Additionally, irrigation also has a few other uses in crop production, which include protecting plants against frost, suppressing weed growth in grain fields and preventing soil consolidation.Irrigation is often studied together with drainage, which is the natural or artificial removal of surface and sub-surface water from a given area. water is required for agriculture. sometimes this water requirement is fulfilled by rain, but there are some dry areas where irrigation is the only process by which water is supplied to crops.

*Irrigation and Agricultural Drainage Engineering*informs students in the application of engineering principles to upkeep useful plant life, with minimum degradation of soil and water resources. The primary objective is to understand soil, water and plant relationships and how they can be applied to better manage natural resources in the production of food and fiber. Aspects covered include: management and maintenance; drainage application and design; and adverse impacts on the environment. This work is of particular value to university students as well as professionals within drainage development, engineering and management.

Table of Contents

1

CHAPTER

Effect of Micro-Irrigation Type, N-Source and Mulching on Nitrous Oxide Emissions in a Semi-Arid Climate: An Assessment Across Two Years in a Merlot Grape Vineyard

Mesfin M. Fentabil[a], Craig F. Nichol[a,*], Gerry H. Neilsen[b], Kirsten D. Hannam[c] ,Denise Neilsen[b], Tom A. Forge[b], Melanie D. Jonesc[a]

[a]*Department of Earth and Environmental Sciences and Physical Geography, University of British Columbia Okanagan, Kelowna, British Columbia V1V 1V7,Canada*

[b]*Agriculture and Agri-food Canada, Summerland Research and Devlopment Centre, Summerland, British Columbia V0H 1Z0, Canada*

[c]*Department of Biology, University of British Columbia Okanagan, Kelowna, British Columbia V1V 1V7, Canadaa*

ABSTRACT

Micro-irrigation, fertigation, and mulching have been proposed to improve the nutrient and water-use efficiency of crop production. The effect of these management practices on the emission of nitrous oxide (N_2O) from vineyards is not well understood and most prior studies were short-term (<1 year). To investigate longer-term effects, a study was conducted in grape (*Vitus vinifera* L. cv. Merlot planted in a sandy loam soil in British Columbia, Canada. The experiment was a factorial treatment design consisting of two micro-irrigation types (Drip or Micro-sprinkler), two nitrogen sources (surface applied Compost or fertigated Urea at a rate of 40 kg N ha^{-1}), and two vineyard floor managements (bark Mulch or "Clean"—meaning bare soil). Frequent measurements of N_2O flux and soil and environmental variables were made over two complete years (2013 and 2014). A considerable portion (37% in 2013 and 61% in 2014) of the annual cumulative N_2O emission (ΣN_2O) occurred during the pre-growing season particularly within the thaw period. In 2013, the annual area-scaled ΣN_2O emissions for Drip was ≈1.8 × Micro-sprinkler, Urea was ≈1.5 × Compost and Clean was ≈1.7 × Mulch. In 2014, ΣN_2O emissions were over 14% higher, likely due to more freeze–thaw events, higher soil mineral N availability (47% higher), but treatments differences were not significantly different. Analysed over two years, micro-sprinkler reduced growing season emissions by 29% and surface application of bark mulch decreased annual area-scaled and yield-scaled ΣN_2O emissions by 28% and 23%, respectively, suggesting bark mulch as a strategy for mitigating N_2O emission. The observed interannual variability in the total N_2O emissions suggests that at least a minimum of 2 years of continuous study may be required to estimate representative annual N_2O emission budgets and to recommend N_2O mitigation strategies in vineyard systems.

Keywords: Nitrous oxide emission; Microirrigation; Drip; Microsprinkler; Mulch; Compost

1. INTRODUCTION

In order to track and manage national and global emissions effectively, N_2O emissions in perennial systems under different management regimes must be quantified across a range of soil types and geographic regions (Carlisle et al., 2010 and Garland et al., 2011). There are limited N_2O emissions data from vineyards (Carlisle et al., 2010 and Steenwerth and Belina, 2010), yet these systems cover more than 7.1 million ha worldwide (FAO, 2013). Vineyards in the semi-arid Okanagan region of British Columbia alone cover approximately 4150 ha (BCGA, 2014). Many vineyards have converted from overhead sprinklers to under vine micro-irrigation (micro-sprinkler or drip) to improve water-use efficiency. The effect of this water management change on N_2O emissions is not known, nor has a preferable micro-irrigation strategy been identified.

Few studies have investigated water management effects on N_2O emission from perennial cropping systems, and none are in a northern climate that experiences a spring ground thaw. Sanchez-Martin et al. (2008) found that drip irrigation resulted in N_2O emissions 70% lower than furrow irrigation in a loamy soil texture after incorporation of broadcast ammonium sulphate in the soil. Sanchez-Martin et al. (2010) observed reduction of N_2O by drip irrigation (28%) compared to furrow irrigation in a sandy clay loam texture fertilized with digested pig slurry, $Ca(H_2PO_4)_2$ and K_2SO_4. Emissions of N_2O from the two previous experiments monitored for a period of less than 5 months and only during the growing season under a Mediterranean climate in Madrid. In California under a Mediterranean climate, Smart et al. (2011) compared emissions from an almond orchard irrigated by stationary fanjet sprinklers and drip irrigation; their observations covered a period of a month in the fall and another month in spring and they reported lower emissions of N_2O from sandy loam soil when mineral N was fertigated via the stationary fanjet sprinklers. In a grape vineyard in California, Suddick et al. (2011)indicated lower N_2O emission for sub-surface drip irrigation than surface drip following a four-day fertigation event. All of the above studies used non-steady-state manual chambers for measuring N_2O emissions and most of them were short-term, lasting few days or months during the growing season. Substantial losses of N_2O have been reported during spring thaw events for other cropping systems in climates with winter temperatures below freezing (Lemke et al., 1999, Müller et al., 2002, Holst et al., 2008 and Tatti et al., 2014). Nevertheless, continuous N_2O emissions data across seasons within a year and between years are lacking from prior studies of woody perennials. Consequently, the long-term effects of various water management strategies on N_2O emissions are still unknown.

Woody perennial crops can be mulched in order to deter weed growth, moderate soil temperature and moisture, reduce evapotranspiration losses, and buffer soil pH in a range optimal to fruit growth (Forge et al., 2013). Organic mulch adds carbon to the soil and depending on the source, can increase the supply of macronutrients to the soil (N–P–K) (Paul 2013) while reducing populations of root-lesion nematodes (Forge et al., 2008) and improving soil biological activity (Neilsen et al., 2014). The addition of compost as a replacement for mineral N fertilizer can also increase soil organic carbon, and supply micronutrients. In prior studies, the addition of organic matter has had contradicting effects on N_2O emission. Addition of organic matter in some field studies resulted in decrease of N_2O emission (Lopez-Fernandez et al., 2007, Livesley et al., 2010,Sanchez-Martin et al., 2010, Steenwerth and Belina, 2010 and Nyamadzawo et al., 2014) while it also resulted in increase of N_2O emission in other studies (Laidlaw, 1993,Cochran et al., 1997 and Pelster et al., 2012). Chantigny et al. (2010) and Pelster et al. (2012) suggested that denitrification rates in sandy loam soils are more limited by C availability than mineral N concentrations, and suggested that the addition of C, through organic amendments such as poultry manure, liquid cattle manure, or liquid swine manure, could lead to higher N_2O emission compared to mineral fertilizers such as calcium nitrate and urea. However, labile organic C also helps in completing the denitrification process, resulting in most of the N_2O converting to N_2, unless there are some inhibiting factors. Such factors include high EC, high pH (Ruiz-Romero et al., 2009), acetylene and sulphides (Knowles, 1982) are present in some organic amendments like manure (Brown, 2012). Most previous studies have used manure as a source of organic matter and little is known about the effect of shredded bark and wood mulch or compost amendments on N_2O emissions.

The effect of management practices such as irrigation or soil amendment on grape yields and N_2O emissions are commonly expressed with respect to area. Expressing N_2O emissions per unit of yield can account for both of these management impacts and may provide a useful metric for greenhouse gas inventories (Venterea et al., 2011). Yield-scaled emissions are reported in other cropping systems (Halvorson et al., 2010, Wei et al., 2010, Gagnon et al., 2011 and Venterea et al., 2011), but reports of yield-scaled N_2O emissions in woody perennial crops are rare (Schellenberg et al., 2012). Additional data on yield-scaled N_2O emission is required to evaluate the viability of any N_2O mitigation strategy without compromising productivity.

This study aimed to determine the effects of agricultural management practices on N_2O emission in a vineyard under a semi-arid climate. The paper presents data collected over two full years, and includes measurements prior to, during and after the growing season. The study investigates how N_2O emissions are affected by: (1) type of micro-irrigation system (Drip vs Micro-sprinkler); (2) nitrogen sources (surface applied Compost or Urea applied through fertigation); and (3) vineyard floor management (shredded bark and wood Mulch versus bare soil, here in referred to as "Clean").

2. MATERIALS AND METHODS

2.1. Study site and experimental design

The study was conducted in the south-central Okanagan Valley (Lat. 49°33′59″N and Long. 119°38′12″W), at the Summerland Research and Development Centre (SRDC) in Summerland, British Columbia, Canada. The site has a 30-year average (1981–2010) annual precipitation of 346 mm, and daily average temperature of 9.6 °C with a minimum daily average of −1.5 °C in January and a maximum daily average of 28.4 °C in July (Environment Canada, 2014a and Environment Canada, 2014b). In this cold-climate wine-growing producing region, below freezing soil temperatures the winter period occur between November and February. Daily maximums in July and August can reach 35–38 °C. The soil is classified as a Skaha Sandy Loam, which is fluvio-glacial in origin, has moderate water holding capacity (Wittneben, 1986) and a cation exchange capacity (CEC) of

10.6 meq/100 g, a pH of 7.4, and a C:N ratio of 8.5 before treatment initiation. Abundant coarse fragments with coatings of secondary carbonates were found at 30–50 cm depth. Secondary carbonates are carbonates ($CaCO_3$ and/or dolomite ($CaMg(CO_3)_2$ that precipitated from the soil solution rather than inherited from a soil parent material. Secondary carbonates can be formed by irrigation with carbonate rich water (Hannam et al., 2015).

The experiment was established on a moderate, south/south-west-facing slope (Fig. 1). The Merlot (*Vitis vinifera*) vines on S04 rootstock were planted in 14 rows of vines in May 2011. The outer rows on each end of the planting were 'guard rows', used to eliminate edge effects and were not directly involved in the experiment. The planting was divided into six blocks consisting of two rows of vines (12 rows total). Each row consisted of five plots and there were five vines per plot. The outer vines were 'guard vines' and middle three vines were 'experimental vines'. Vine spacing was 1.2 m within rows and 3.0 m between rows. Vine density was 2, 778 vines/ha. A common orchard grass mix was grown in the 1.5 m wide inter-row (alley). Plots were kept weed-free via the use of herbicides (primarily glyphosate).

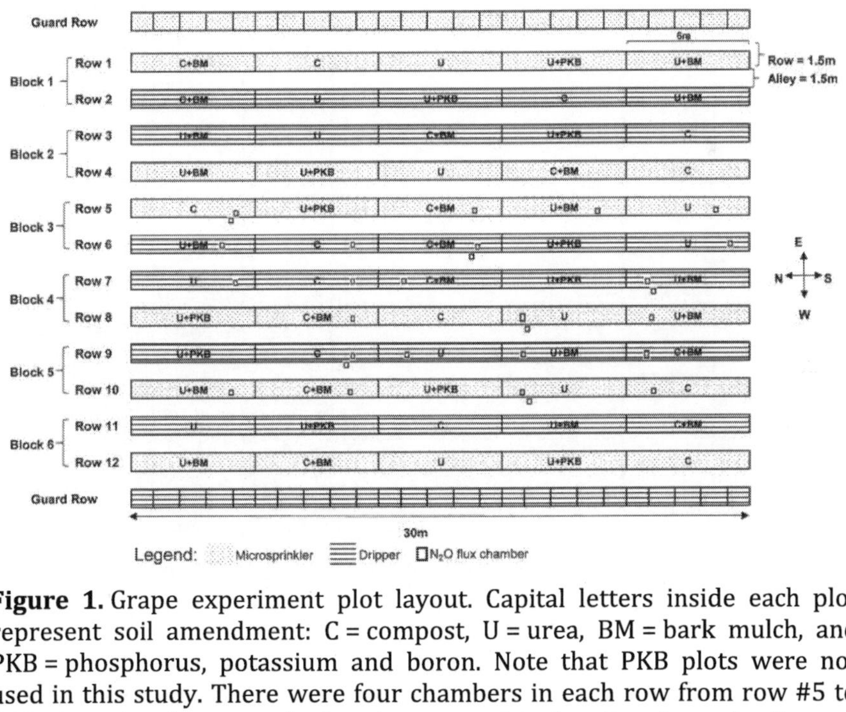

Figure 1. Grape experiment plot layout. Capital letters inside each plot represent soil amendment: C = compost, U = urea, BM = bark mulch, and PKB = phosphorus, potassium and boron. Note that PKB plots were not used in this study. There were four chambers in each row from row #5 to #10. One chamber was deployed at a random location in each alleyway between rows #5 to #11.

Experiments were factorial treatment designs consisting of two micro-irrigation types (Drip or Micro-sprinkler), two nitrogen sources (Compost or Urea), and two vineyard floor managements (bark Mulch or Clean). Main plots (complete rows) were comprised of two different low pressure irrigation systems (Drip, Micro-sprinkler) whereas subplots consisted of five fertilization/amendment combinations planted in each of the six replicates for each irrigation treatment. The Compost plots and Urea plots had different irrigation lines to enable selective fertigation of urea plots. Two 4 L h^{-1} pressure compensating drippers were placed on opposite sides of each vine along the vine row (Fig. 2A) in Drip plots, and one 20 L h^{-1} flow regulating micro-sprinkler (AquaSmart 2000 Naandanjain Ltd., Israel) was placed mid-way between two adjacent vines in Micro-sprinkler plots (Fig. 2B). Both micro-irrigation types were suspended about 0.3 m above the soil surface. An atmometer (ETGage Co., Loveland, CO) was used to estimate potential evapotranspiration, which was combined with a crop coefficient model (Neilsen et al., 2015) to estimate water required to replace 100% of the previous day's losses to evapotranspiration via four equal applications every 6 h over a 24 h period controlled by a CR10X datalogger (Campbell Scientific, Logan, UT). The irrigation season extended from May through October in both 2013 and 2014.

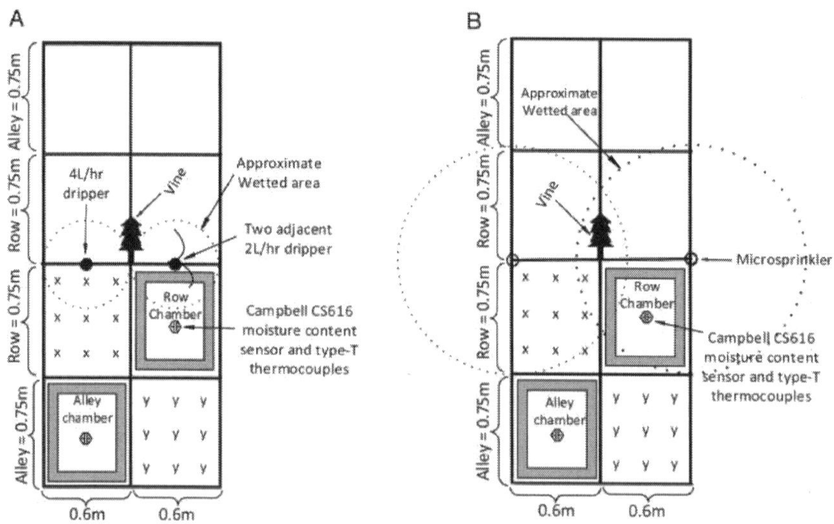

Figure 2. Spatial representation of a single vine functional unit in: (A) Drip irrigated plot, and (B) Micro-sprinkler irrigated plot. *"x"* and *"y"* represent examples of soil sampling locations for row chamber and alley chamber for one sampling event, respectively. Similar locations relative to other experimental vines (not shown) in the same plot were sampled for each soil sampling event in May, June, August, and October in 2013 and every month from April to August, and October in 2014.

All the grape plots received a total of 40 kg N ha^{-1} in the form of either urea or compost. Urea was fertigated for four weeks from mid-May to mid-June from 2011 to 2013, and from late May to late June in 2014 timed to coincide with grape development. Compost and Mulch made from shredded bark and wood-chips were applied on a 1.5 m wide strip centered on the vine row. Compost was used as a nitrogen source in the plots that were not fertigated with Urea. Mulch was used primarily to minimize evaporation of water from the soil. Both compost and mulch also add carbon to the soil. Compost was applied in late May each year (2011–2014) and was manually incorporated into the top 5 cm of soil. The locally produced compost was made from approximately 15% grape pomace, 20% straw, 25% shredded bark and wood chips and 40% cow manure and screened through 0.6 cm mesh. Prior to application in the field, the N content of compost was determined on air-dried and ground samples using a LECO FP-528 (Leco Corporation, St. Joseph, MI). Based on the C/N ratio of the compost (3 year average = 20), we estimated that first-year mineralization of N would be no more than 20% of the total N content (Gale et al., 2006). This liberal estimate of Potentially Available N (PAN) was used to adjust the compost application rate each year to provide an estimated 40 kg PAN/ha. The mineralization of N from surface-applied compost in the second and third years after application is not well understood. Consequently, we did not attempt to account for previous applications when making the calculations for any given year. While we used a liberal estimate of PAN, this approach nonetheless creates the potential for accumulation of residual mineralizable N, and the possibility that in later years of the experiment the release of mineral N from the compost exceeded the 40 kg N/ha target.

Mulch applied to the row was composed of shredded bark and wood-chips (primarily from *Pinus contorta* var *latifolia* and *Picea glauca*) generated as waste from local sawmills. It was surface applied in late May of every second year (2011, 2013) to maintain a total mulch depth of approximately 10 cm. New mulch was added over top of existing mulch without disturbance to the underlying material. Mulch was not applied in 2012 and 2014 because it was still sufficiently thick to suppress weeds.

2.2. Soil sampling and analyses

At each soil sampling time, soil cores to a depth of 15 cm were collected from nine locations (Fig. 2) in the row and another nine location the alley using a 2-cm diameter auger. Samples were composited to get a representative sample for the row and the alley part of each plot. Soil sampling was conducted in May, June, August, and October in 2013, and every month from April to August, and in October in 2014 ($n = 10$). Soil samples were kept frozen until extraction and analysis. Field moist soils were extracted for exchangeable nitrate-N and nitrite-N (NO_3^-–N) and ammonium-N (NH_4^+–N) with 2 M KCl using a 1:5 soil to extractant ratio and a 1-h shaking time followed by filtration through Whatman No. 40 filter paper. Extracts were frozen at −20 °C and defrosted overnight prior to analysis for NO_3^-–N and NH_4^+–N using a segmented flow analyzer (SFA, Model 305D, Astoria Pacific International, Clackamas, OR). Salt-extractable organic carbon (SEOC) samples were extracted similarly (in 2 M KCl, 1:5 soil extractant ratio), and filtered through a 0.45-μm membrane filter (Millipore Corp, USA) prior to storage at −20 °C (Chantigny et al., 2008). The SEOC in the extracts were analysed by Aurora 1030W OI Analytical TOC analyzer (OI Analytical, USA). Calibration standards were made from potassium hydrogen phthalate in 2 M KCl. Calibration standards and samples were diluted 1:12 with ultrapure water to keep the chloride level to approximately 1% to prevent interference effects. 2 ml of the sample were reacted with 2 ml of 5% phosphoric acid and 2 ml of 10% sodium persulphate at 98 °C to oxidize and liberate the organic carbon and measure it as carbon dioxide.

Total C and N content were measured by combusting 15–20 mg finely ground air dried sample using Costech 4010 Elemental Analyzer with thermal conductivity detection. CEC was measured using an ammonium acetate extraction buffered to pH 7. Soil pH and EC were measured by pH-meter (WTW inoLab pH 7200) and EC-meter (WTW inoLab Cond 7200) on an extract from 1:2 (w/v) deionized water to finely ground air-dried soil which was sifted through a 2.00 mm sieve. Additional soil sampling and nutrient availability assessment using soil and ion exchange resin sampling is reported inHannam et al. (2016).

In-situ volumetric soil water content (0–30 cm depth) and temperature (at 2 cm and 10 cm depth) for each plot were measured continuously at 1-h intervals using 30 cm length time domain reflectometry (TDR) probes installed vertically (Campbell Scientific, CS616) and type-T thermocouples, respectively, monitored by Campbell Scientific CR1000 data logger. All TDR probes and thermocouples were installed permanently at the center of each plot where N_2O flux was measured and kept undisturbed throughout the experiment. Soil bulk density samples were collected at 15 cm and 30 cm away from the dripper/micro-sprinkler towards the alley to represent the fertilized strip, and at the center of the alley, from samples from 3 to 9 cm depth using 6 cm diameter copper collars that were driven into the soil using a soil bulk density sampler. An estimate of the fraction of water filled pore space (WFPS) was calculated as:

$$\mathrm{WFPS} = (\theta_{\mathrm{v}}/[1-(\rho_{\mathrm{b}}/\rho_{\mathrm{p}})]) \tag{1}$$

where θ_v is the volumetric water content of the soil ($cm^3\ cm^{-3}$), ρ_b is the bulk density of the soil ($g\ cm^{-3}$) and ρ_p is the approximate mineral particle density ($2.65\ g\ cm^{-3}$). Measurement of volumetric water content was conducted in the growing season as well as winter period. However, during winter the Campbell CS616 sensors measured the portion of water in soil that existed only in liquid phase (i.e., excluding the ice phase portion of water in the soil). Therefore, the WFPS during winter were likely underestimated. Thus, only the WFPS data collected during the growing season of each year were used to determine the statistical differences in WFPS among treatments.

2.3. N_2O flux measurements

Emissions of N_2O were measured in three replicates at the three middle blocks (from block 3 to 5) comprising all soil amendment except the treatment containing urea plus phosphorus, potassium and boron (PKB) (Fig. 1); replicates in other blocks were dedicated for parallel studies of carbon isotopes in carbon dioxide emissions. The gas flux sampling strategy was designed to capture high N_2O emissions over short time periods caused by triggering events such as fertilization, irrigation, mulching, and major weather events such as spring thaw and intensive rainfall. The regular monitoring schedule involved sampling twice a week during fertigation, once a week during irrigation, and once every second week in winter. Sampling frequency was increased to two or three times a week around short-term weather and management events such as compost and mulch application, irrigation initiation, spring thaw or intensive rainfall. Regular flux monitoring started in January 2013. There were a total of 48 rounds of sampling each year, in 2013 and 2014. Fluxes were measured using non-flow through non-steady-state (NFT-NSS) chambers (Rochette and Eriksen-Hamel, 2007). Rectangular stainless steel frames (0.68 m × 0.56 m × 0.15 m depth) were installed to a depth of 0.13 m and left undisturbed for the duration of the experiment. At sampling, an insulated vented lid (with top closed, headspace ~0.11 m) was tightly fitted to the frames. The size and location of the chamber within the plot were chosen to capture a representative emission footprint by taking into account the width of the alley and row (fertilized strip), the location of vines, and the drippers/micro-sprinklers (Fig. 2). To prevent the distortion of soil water distribution that may be caused by placement of the chamber collar, the $4\ L\ h^{-1}$ drippers adjacent to each chamber edge were replaced by two half-capacity smaller drippers ($2\ L\ h^{-1}$), one dripping into the chamber at the inside edge and the other dripping on the outside edge of the chamber hence in the adjacent "quadrant" on the opposite side of the row (Fig. 2A).

Gas samples were collected using pre-evacuated 12 ml double-wadded Exetainers (Part No: 737W, Labco Ltd.) that were evacuated to 200 mTorr, flushed with helium, and re-evacuated to 200 mTorr. During sampling, the Exetainers were over-pressurized by injecting 20 ml of sample using a 25-ml gas-tight glass syringe (catalog No. 03 378 207, National Scientific, Rockwood, TN). Samples were taken at 0, 7, 14, and 21 min in the summer when the magnitude of N_2O flux was usually high. Nitrous oxide flux was usually low in winter and hence chamber deployment duration and sampling intervals were extended during winter to allow build up enough gas concentrations in the chamber headspace. Samples were taken at 0, 10, 20, and 30 min in the winter. The samples were accompanied by three low and three high field standards (containing 0.220 ppm and 2.20 ppm N_2O–N) which were prepared on the sampling day. Nitrous oxide concentrations were analyzed within one week of sampling date using a gas chromatographs (Bruker 456 GC, Bruker) equipped with an electron capture detector (ECD) and a CTC Combi-Pal auto sampler (CTC Analytics AG, Zwingen, Switzerland). The soil-surface N_2O fluxes were determined by calculating slope (dC/dt) of concentration (C) vs time (t) at chamber closer ($t = 0$) via either linear or non-linear regression (as appropriate) using the equations of Rochette and Hutchinson (2005). Non-linear regression (Hutchinson and Livingston, 1993) was used when the accumulation of N_2O decreased with time. Linear regression was used when the accumulation of N_2O was consistent with time (Rochette and Eriksen-Hamel, 2007). Climate data (relative humidity, air temperature, and pressure) for flux calculations were obtained from a weather station located within 0.5 km (Environment Canada, 2014b).

Treatments were applied on the row functional unit only (Fig. 2) and hence unless mentioned otherwise treatment comparison was based on measurements of N_2O emission from the row. The alley functional unit did not receive treatment and flux was measured by six chambers deployed at six alleys. The N_2O emission measurements from alley chambers were only used to calculate the emission factors as discussed below.

Vineyard-scaled emissions factors (EFs), uncorrected for background emission for each treatment, were expressed as the percentage of the applied N emitted as N_2O–N using:

$$\mathrm{EF} = \left(\sum \mathrm{N_2O_{weighted}} / \text{Applied available } \mathrm{N_{source}}\right) \times 100\% \tag{2}$$

where $\sum N_2O_{weighted}$ is the area weighted annual cumulative N_2O emission of a field (kg N_2O–N ha^{-1}) calculated by weighting row and alley fluxes by the area covered by the row and alley (each accounting for 50% of the field);

and Applied available $N_{source}$is the application rate of N fertilizer (40 kg N ha^{-1}).

2.4. Yield-scaled N_2O emissions

Harvesting occurred on October 29, 2013 and October 20, 2014. Grape yield was determined from three central vines in every plot and extrapolated to a per hectare yield (kg ha^{-1}). Grape yield and fruit quality are a subject of separate future publication and are not presented in this paper; but grape yield was used in the present paper to calculate yield-scaled N_2O emissions (g N Mg^{-1}) for the treatments by dividing the annual commutative N_2O emissions (g N ha^{-1} $year^{-1}$) by grape yield (Mg ha^{-1} $year^{-1}$).

2.5. Data and statistical analysis

Data were analysed to capture fluxes during (i) the spring thaw in the pre-growing season (PreGS) (January through April); (ii) the growing season (GS) (May through October); (iii) the period between harvest and the first ground thaw (post-harvest or post-growing season [PGS]: November and December); and (iv) annually (January through December). Area-scaled $\sum N_2O$ for individual plots were calculated using linear interpolation of flux rates between sampling days (Millar et al., 2012) and by extrapolation to per hectare of fertilized strip. To correct for both lack of normality and nonconstancy of the error variance, all data (except pH and WFPS in 2013 and, pH and $\sum N_2O$ in 2014) were first transformed using Box–Cox transformations (Box and Cox, 1964) using the SAS transgress procedure, but all data are reported as untransformed means. Effects of treatments on N_2O emissions and environmental variables were determined using the PROC MIXED procedure in SAS, with repeated measurements in the model option (Version 9.3; SAS® Institute, Inc., Cary, NC) and block, and block-by-irrigation treated as random effect. Model selection was conducted as described by Littell et al. (2006). Since N_2O emissions were measured over unequally spaced time intervals, a spatial power law (SP(POW)) covariance structure was used. The SP(POW) structure for unequally spaced longitudinal measurements provides a direct generalization of the autoregressive model (order one) for equally spaced measurements. The SP(POW) models the covariance between two measurements at times t_1 and t_2 as:

$$\mathrm{Cov}\left[Y_{t_1}, Y_{t_2}\right] = \sigma^2 \rho^{|t_1 - t_2|} \tag{3}$$

where Y is a measurement, ρ is an autoregressive parameter assumed to satisfy $|\rho| < 1$ and σ^2 is an overall variance.

Similarly, the effects of treatments on $\sum N_2O$ were conducted using the PROC MIXED procedure, for individual years, and for the 2-year combined data with years as repeated measures. Contrary to daily measurements, cumulative measurements were equally spaced in time (i.e., repeated over a year) and compound symmetry covariance structure was used in the "type" statements of the repeated model to draw overall statistical comparisons among management factors imposed over two years. Pairwise means comparisons were performed using the PDIFF statement and Tukey–Kramer adjustment method. Unless otherwise mentioned, a p-value <0.05 was used for fixed effects and means separation.

3. RESULTS

3.1. Weather, soil WFPS and temperature

Environmental conditions in the two experimental years were typical of the region. The mean daily air temperature in 2014 (9.8 °C) was slightly higher than 2013 (9.6 °C) which was the same as the 30-year average (9.6 °C) (Table 1). Higher total precipitation with higher proportion during the fertigation period occurred in 2013. The largest individual rainfall events occurred in 2013, where three events occurred (June 18, 12 mm; June 20, 26 mm; June 24, 32 mm). In 2014, a single event of 35 mm was noted on June 13. The total irrigation applied in both years were similar.

Table 1. Summary of climate and soil data during experimental years 2013 and 2014 including thaw dates, precipitation, irrigation, average air and soil temperatures. Data presented during the pre-growing season (PreGS, January through April), the growing season (GS, May through October), and post-growing season (PGS, November and December).

Parameter	2013			2014		
	PreGS	GS	PGS	PreGS	GS	PGS
Date of first thaw	January 9th	–	–	January 10th	–	–
Date of last thaw	Febuary 20th	–	–	March 4th	–	–
Total precipitation (mm)[a]	38/55	137/104	36	22/43	74/116	71
Total irrigation (mm)	–	31/241	–	–	46/236	–
Total water inputs[x]	93	513	36	65	472	71
Average soil temperature[y] (°C)	3.1	17.5	0.6	3.7	18.2	1.6
Average air temperature (°C)	3.5	17.2	−1.1	2.6	17.5	0.8
No of days soil temperature <0 °C	36	0	36	40	0	20
No of days air temperature <0 °C	29	0	37	40	0	23
No of days soil temperature <−5 °C	7	0	1	0	0	2
No of days air temperature <−5 °C	6	0	12	14	0	9

[x] Total water inputs = Total precipitation + Total irrigation.
[y] Soil temperature was measured at 2 cm depth.
[a] Data separated by "/" indicates either thaw period/rest of PreGS totals or fertigation period/irrigation period totals.

The freeze–thaw periods of the pre-GS of 2013 and 2014 had different characteristics (Table 1, Fig. 3). The thaw period in the pre-GS of 2013 was shorter (Table 1). Average daily air temperature increased linearly from below to above zero, followed by soil temperature. During the thaw period in 2014, average daily air temperatures fluctuated above and below zero three times in cycles lasting several days prior to staying above zero. There were more days in the PreGS of 2014 when air and soil temperature were below 0 °C.

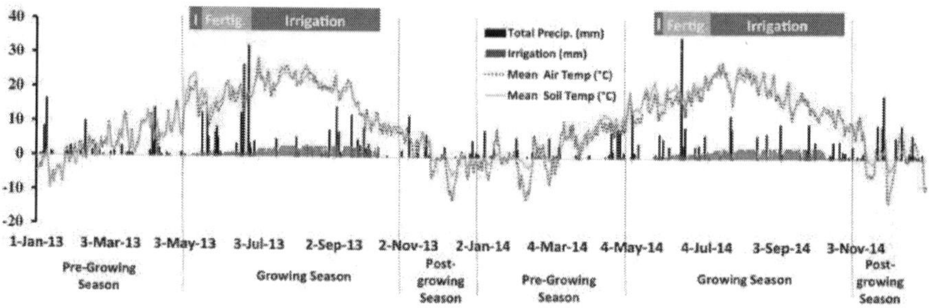

Figure 3. Air and soil temperature, precipitation and irrigation inputs for 2013 and 2014. "Fertig" in the top insert represent periods of fertigation, the application of nitrogen through the irrigation system.

Soil temperatures varied among treatments during the growing season but the mean differences were very small, and hence not visually differentiable when plotted (Fig. 4). Near surface soil temperature was not affected by irrigation type while statistical differences were found for different N-sources ($P < 0.05$) and floor management types ($P < 0.001$). However, these differences were likely too small (<0.6 °C) to drive any substantive changes in N_2O emissions among treatments.

Treatment and year did not affect soil bulk density. Two years after establishment of the treatments, soil bulk density remained similar across the vineyard averaging 1.31 g cm^{-3}($n = 54$, two replicates per plot and one replicate per alley). These data were used to calculate WFPS.

During the GS of both years, WFPS varied significantly by irrigation ($p < 0.01$) and floor management ($p < 0.01$) but not by N-source (Fig. 4). In 2013, the GS average WFPS for Drip (51.6%) > Micro-sprinkler (45.4%) and Mulch (51.6%) > Clean (45.4%) which was repeated in 2014 when the GS average WFPS for Drip (53.8%) > Micro-sprinkler (41.7%), and Mulch (50.4%) > Clean (45.1%). In both years, WFPS started to vary significantly by irrigation type and floor management approximately one week after irrigation started. However, differences in WFPS were lessened immediately following some isolated intense rainfalls events especially in late May and June of 2013 (Fig. 3 and Fig. 4). The difference in daily GS

WFPS among the treatments was most pronounced in July and August when the highest soil and air temperatures were recorded. Mean GS WFPS in Drip irrigated plots reached maximum of 59% and 64% in 2013 and 2014, respectively. Mean GS WFPS in Mulched plots had maximum of 61% in 2013 and 56% in 2014. Individual measurements in single replicates were higher.

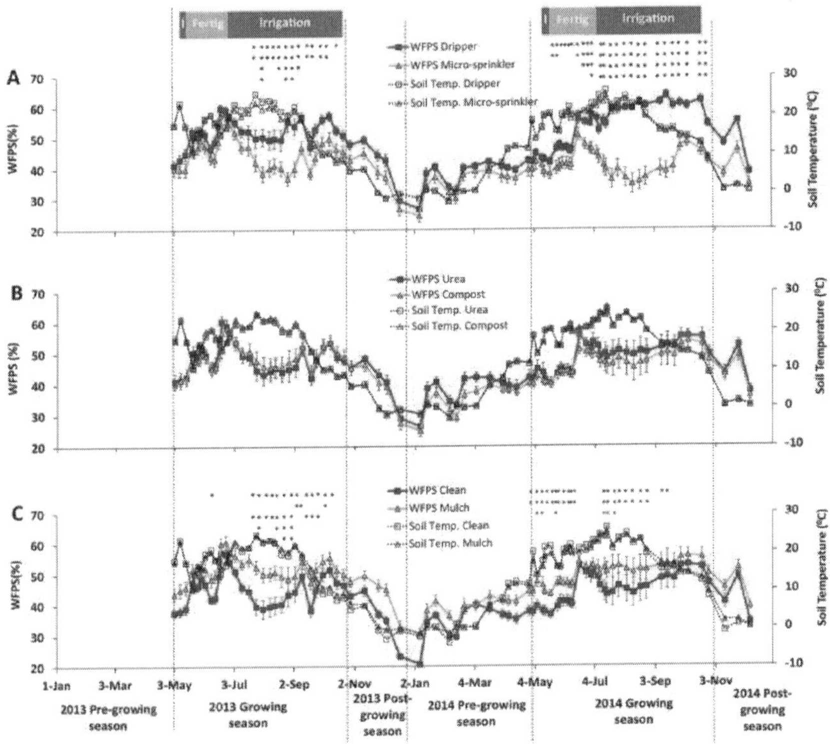

Figure 4. Soil water filled pore space (WFPS) and soil temperature over 2013 and 2014 across: (A) irrigation type (B) nitrogen source (C) orchard floor management. Bars indicate standard error of the mean. *, **, ***, **** indicate differences of least squares means of WFPS using the Tukey–Kramer adjustment, at $p < 0.05$, $p < 0.01$, $p < 0.001$, $p < 0.0001$, respectively. Differences of least squares means of temperature are not shown. "Fertig" in the top insert represent periods of fertigation, the application of nitrogen through the irrigation system.

3.2. Daily N_2O emissions

The temporal patterns of N_2O emissions (Fig. 5) were similar across treatments, from within the alley, and between years. N_2O emission spikes were observed mainly within two periods: during freeze–thaw cycles in the PreGS and during the fertigation period in the GS of each year. Emission during the PGS was negligible. There was only one noticeable difference in the pattern of N_2O emission between years, observed in July and August 2013, when there was a N_2O emission spike under Drip but not under Micro-sprinkler; a similar difference was not observed in 2014.

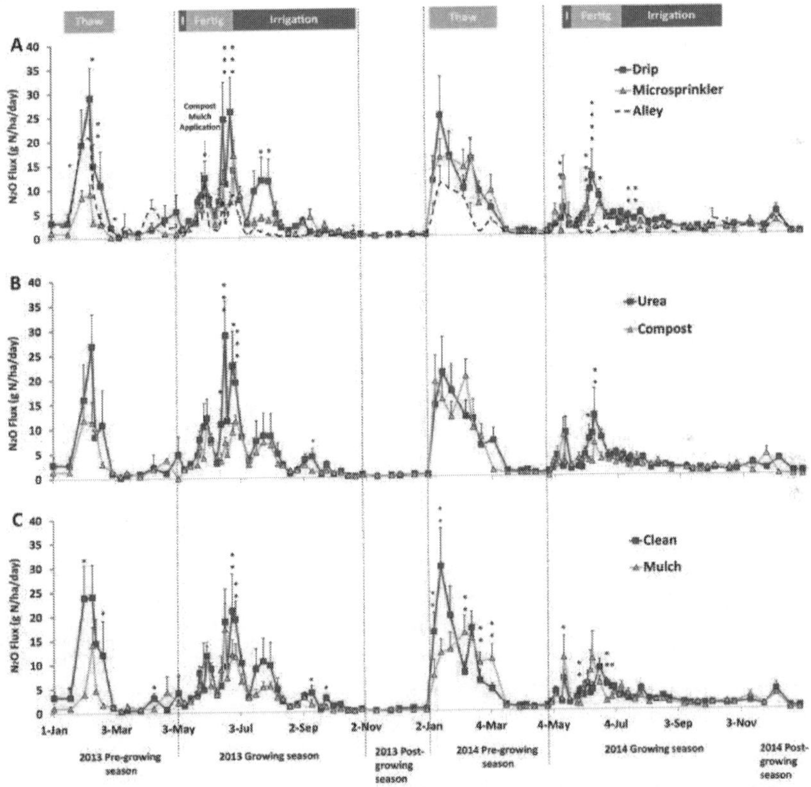

Figure 5. Mean daily measurement of N_2O emissions in 2013 and 2014 as averaged by: (A) micro-irrigation type and alley position (B) nitrogen source (C) orchard floor management. Plus capped bars indicate standard error of the mean ($n = 12$). *, **, ***, **** indicate differences of least squares means using the Tukey–Kramer adjustment, at $p < 0.05$, $p < 0.01$, $p < 0.001$, $p < 0.0001$ respectively. "Fertig" in the top insert represent periods of fertigation, the application of nitrogen through the irrigation system.

Treatment effects on daily N_2O emissions were not consistent across seasons within a year and between years (Table 2). During the PreGS in 2013, Floor management and Date accounted for most of variation in N_2O emissions. This caused lower N_2O emissions in Mulched than Clean plots in 4 of the 12 monitoring days (Fig. 5C). During the 2013 GS, N-source, Floor management and Date accounted for most of the variation in N_2O emissions; N_2O emissions under Micro-sprinkler was significantly lower than Drip in 4 of the 32 monitoring days (Fig. 5A). In addition, N_2O emissions in Compost plots were significantly lower than Urea plots in 4 of the 32 monitoring days (Fig. 5B).

Table 2. *F*-value of repeated measures analysis of daily N_2O emissions during the pre-growing season (PreGS, January through April) and during the growing season (GS, May through October) in 2013 and 2014.

Effect	2013				2014			
	PreGS		GS		PreGS		GS	
	F Value	Num DF	*F* Value	Num DF	*F* Value	Num DF	*F* Value	Num DF
Irrigation (I)	3.86	1	1.15	1	0.64	1	16.42^{*}	1
N-source (N)	0.59	1	5.43^{*}	1	0.48	1	0.00	1
Floor mgmt (F)	7.00^{**}	1	7.35^{**}	1	0.19	1	1.54	1
Dates (D)	33.29^{****}	10	26.10^{****}	31	30.69^{****}	11	8.9^{****}	31
I x D	2.35^{*}	10	2.15^{***}	31	0.56	11	2.21^{***}	31
N x D	0.87	10	1.93^{**}	31	1.31	11	1.11	31
F x D	3.66^{***}	10	1.18	31	3.29^{*}	11	1.45^{*}	31

*, **, ***, **** indicate differences between treatments significant at $p < 0.05$, $p < 0.01$, $p < 0.001$, $p < 0.0001$, respectively. Irrigation occurred from May 8 to October 23 in 2013 and from May 21 to October 23 in 2014. Urea was fertigated from May 14 to June 25 in 2013 and May 28 to July 7 in 2014 while compost was applied in late May in both years. Bark mulch was surface applied late May of only 2013.

In 2014 most the variation in N_2O emissions across seasons were caused by irrigation and sampling date (Table 2). During the PreGS the N_2O emissions in Mulched plots were significantly lower than Urea plots in 5 of the 12 monitoring days (Fig. 5C). During the GS the N_2O emissions under micro-sprinkler irrigation were significantly lower than drip irrigation in 6 of the 32 monitoring days (Fig. 5A). Generally, enhanced N_2O fluxes during the GS of both years were observed in the days when WFPS was usually greater than 45% (for example May 31, June 18, and June 25 in 2013 and June 18 in 2014) (Fig. 4 and Fig. 5).

3.3. Seasonal and annual cumulative N_2O emissions

Treatment effects on $\sum N_2O$ emissions were not consistent across seasons and years (Table 3). For example, in 2013 irrigation type resulted in a significant difference in the PreGS $\sum N_2O$ emissions but did not significantly affect the GS $\sum N_2O$ emissions of same year. Irrigation type did not result in significant differences in the PreGS $\sum N_2O$ emissions of 2014, contrary to 2013. During the PreGS of 2013, area-scaled $\sum N_2O$ emissions of Drip < Micro-sprinkler and Mulch < Clean. During the GS, area-scaled $\sum N_2O$

emissions of Compost < Urea and Mulch < Clean. Overall, the 2013 annual area-scaled $\sum N_2O$ emissions for Drip ≈ 1.8 × Micro-sprinkler ($p = 0.118$), Urea ≈ 1.5 × Compost ($p = 0.027$) and Clean ≈ 1.7 × Mulch ($p = 0.003$) but only the differences from N-source and floor management were significant. None of the simple treatment factors caused significant effects on 2014 seasonal and annual area-scaled $\sum N_2O$ emissions ($p > 0.05$).

Table 3. Effects of irrigation type, nitrogen source and orchard floor management on seasonal and annual N_2O emissions during the pre-growing season (PreGS, January through April) and during the growing season (GS, May through October) in 2013 and 2014.

Effect	N_2O emission[x] (kg ha^{-1}[season or year]$^{-1}$)					
	2013			2014		
	PreGS	GS	Annual	PreGS	GS	Annual
Irrigation (I)						
Sprinkler	0.27 (0.03) a[y]	0.65 (0.09)	0.95 (0.11)	0.92 (0.14)	0.43 (0.06)	1.45 (0.18)
Dripper	0.76 (0.19) b	0.97(0.17)	1.74 (0.34)	0.97(0.14)	0.54(0.05)	1.63 (0.19)
N-source (N)						
Compost	0.41 (0.09)	0.64 (0.08) a	1.06 (0.16) a	0.97 (0.14)	0.46 (0.05)	1.53 (0.18)
Urea	0.61 (0.19)	0.99 (0.17) b	1.63 (0.34) b	0.93 (0.14)	0.52 (0.06)	1.55 (0.19)
Floor mgmt (F)						
Mulch	0.29 (0.08) a	0.68 (0.12) a	0.98 (0.16) a	0.86 (0.13)	0.49 (0.06)	1.45 (0.17)
Clean	0.73 (0.18) b	0.95 (0.15) b	1.71 (0.33) b	1.03 (0.15)	0.49 (0.06)	1.63 (0.20)
Pr > F[z]						
Irrigation (I)	**	ns	ns	ns	ns	ns
N-source (N)	ns	*	*	ns	ns	ns
Floor mgmt (F)	***	*	**	ns	ns	ns
I x N	ns	ns	ns	ns	ns	ns
I x F	ns	ns	ns	ns	ns	ns
N x F	ns	ns	ns	ns	*	ns
I x N x F	ns	ns	ns	*	ns	*

[x] Mean N_2O emission calculations for treatments were based on a 1.5 m wide fertilized strip.
[y] Numbers in parentheses following mean are standard error of the mean, followed by different lowercase letters within columns indicating differences of least squares means using the Tukey-Kramer adjustment, at $p < 0.05$.
[z] *, **, ***, **** and ns indicate that means are significantly different at $p \leq 0.05$, 0.01, 0.001, 0.0001 or are not significantly different, respectively.

The relative contribution of seasons towards the total annual $\sum N_2O$ varied across years (Table 3). In 2013, the majority (62%) of the annual $\sum N_2O$ emissions occurred during the GS, followed by PreGS (37%) and PGS (1%) whereas in 2014 the majority of N_2O emission occurred during the PreGS (61%) followed by GS (32%) and PGS (7%) (Fig. 5and Table 3). The average N_2O emissions in all the treatments in the PreGS of 2014 (0.95 kg ha^{-1}) was much higher than 2013 (0.52 kg ha^{-1}) while the emissions in the GS of 2013 (0.81 kg ha^{-1}) was much higher than 2014 (0.49 kg ha^{-1}). On average, the annual cumulative N_2O emissions in 2014 were over 14% greater than 2013.

The 2-year averaged annual area-scaled $\sum N_2O$ emissions were: Micro sprinkler < Drip (at $p = 0.20$); Compost < Urea (at $p = 0.15$); and Mulch < Clean (at $p = 0.02$) (Table 4). However, only the difference in floor management type was statistically significant. Surface application of bark mulch decreased the 2-year mean $\sum N_2O$ emission by 28%.

Table 4. Effects of irrigation type, nitrogen source and vineyard floor management on 2-year mean seasonal and annual N_2O emissions, 2-year mean N_2O emissions factor (EF, N_2O emissions per unit of total N applied), and 2-year mean yield-scaled N_2O emissions (N_2O: grape yield). Year was included as a repeated measure factor in the ANOVA.

Effect	N_2O emissions (kg ha^{-1} [season or year]$^{-1}$)				N_2O[z] emission factor (%)	Yield-scaled N_2O (g Mg^{-1})
	PreGS	GS	PGS	Annual		
Irrigation (I)						
Sprinkler	0.60(0.08)[x]	0.54 (0.08) a	0.06 (0.01)	1.20 (0.14)	2.53 (0.13)	51.6 (6.3)
Dripper	0.86(0.17)	0.76 (0.11) b	0.07 (0.01)	1.69 (0.27)	3.14 (0.24)	54.9 (5.8)
N-source (N)						
Compost	0.69(0.11)	0.55 (0.07)	0.06(0.02)	1.30(0.17)	2.65(0.15)	50.1(4.6)
Urea	0.77(0.17)	0.75 (0.12)	0.06(0.01)	1.59(0.27)	3.02(0.24)	56.5(7.2)
Floor mgmt (F)						
Mulch	0.58(0.10) a	0.58(0.09)	0.06(0.01)	1.21(0.17) a	2.55(0.15) a	46.3(5.7) a
Clean	0.88(0.16) b	0.72(0.11)	0.06(0.01)	1.67(0.26) b	3.12(0.24) b	60.3(6.1) b
Pr > F[y]						
Irrigation	ns	**	ns	ns	ns	ns
N-source	ns	ns	ns	ns	ns	ns
Floor mgmt	**	ns	ns	*	*	*
Year	***	***	****	*	*	ns

[x] Numbers in parentheses following mean are standard error of the mean, followed by different lowercase letters within columns indicating differences of least squares means using the Tukey–Kramer adjustment, at $p < 0.05$.
[y] *, **, ***, **** and ns indicate that means are significantly different at $p \leq 0.05, 0.01, 0.001, 0.0001$ or are not significantly different, respectively.
[z] N_2O emission factor calculations for treatments were based on the N_2O emissions measured at both the row and the alley functional units.

The effects of treatments on seasonal area-scaled $\sum N_2O$ emission were somewhat different from the annual $\sum N_2O$ emission over the two year period (Table 4). During the GS, irrigation treatment resulted lower area-scaled $\sum N_2O$ emissions in Micro-sprinkler than Drip. Floor management treatment caused lower area-scaled $\sum N_2O$ emissions in Mulch than Clean only in the PreGS. As the majority (49%) of N_2O emission over the two years occurred during the PreGS (when there was no irrigation), only floor management effects were significant when all (preGS, GS, PGS) seasonal $\sum N_2O$ were added together. Hence, there was lower area-scaled $\sum N_2O$ in Mulch than Clean over the two years period.

Emission factors (N_2O emissions per unit of total N applied uncorrected for background emission) ranged between 2.53% to 3.14% across treatments (Table 4). All treatments received an estimated available N-rate (40 kg ha^{-1} $year^{-1}$) which means the emission factors were affected in a similar way as the annual area-scaled $\sum N_2O$ emissions. There were no significant differences in the emission factors caused by irrigation type or N-source. However, floor management caused significant difference; the emission factor for Mulch was lower than Clean ($p < 0.05$). Yield-scaled $\sum N_2O$ emissions for Mulched plots were lower than Clean plots by 23%. Hence, under our experimental conditions, production of an equivalent amount of grape yield using mulch would generate lower N_2O than production without mulching.

3.4. Soil nutrients and chemistry

In both years, soil NO_3^-–N concentrations were not affected by irrigation or N-source (Table 5). Soil NO_3^-–N concentrations were significantly affected by floor management as soil NO_3^-–N concentrations in Clean plots were 4 times higher than Mulched plots ($p < 0.0001$). On average, soil NO_3^-–N concentrations increased by 58% from 2013 to 2014.

Table 5. Effects of irrigation type, nitrogen source and vineyard floor management on mean extractable NO_3^--N, NH_4^+- N, salt-extractable organic carbon (SEOC), pH and electrical conductivity (EC) of soil (0–15 cm depth) during the growing season (May through October) in 2013 ($n = 96$) and 2014 ($n = 144$) (April through October).

Year	Effect	NO_3^-–N ($mg\ kg^{-1}$)	NH_4^+–N ($mg\ kg^{-1}$)	SEOC ($mg\ kg^{-1}$)	pH	EC ($\mu S\ cm^{-1}$)
2013	Irrigation (I)					
	Sprinkler	22.1(2.2)[x]	6.6(0.5)	398.6(23.3)	7.15(0.02)	195.9(7.2)
	Dripper	37.8(5.9)	6.8(0.7)	497.5(33.4)	7.14(0.03)	246.2(11.1)
	N-source (N)					
	Compost	28.2(3.8)	6.8(0.6)	546.4(35.1) a	7.20(0.03) a	241.3(9.0) a
	Urea	31.7(5.3)	6.5(0.5)	349.8(14.8) b	7.09(0.02) b	200.7(9.9) b
	Floor mgmt (F)					
	Mulch	12.4(1.3) a	7.2(0.6)	490.2(35.8) a	7.15(0.02)	174.6(6.3) a
	Clean	47.6(5.3) b	6.1(0.6)	405.9(19.9) b	7.14(0.02)	267.5(9.5) b
	Pr > F[y]					
	Irrigation (I)	ns	ns	ns	ns	ns
	N-source (N)	ns	ns	****	***	****
	Floor mgmt (F)	****	ns	*	ns	****
2014	Irrigation (I)					
	Sprinkler	41.1(5.0)	9.9(0.8)	464.7(38.1)	6.90(0.04)	320.1(18.7)
	Dripper	53.8(7.7)	8.3(0.6)	536.0(39.3)	6.91(0.03)	330.2(21.5)
	N-source (N)					
	Compost	43.6(5.1)	10.4(0.8) a	630.9(43.0) a	7.02(0.03) a	340.4(16.7) a
	Urea	51.3(7.6)	7.8(0.6) b	369.8(26.6) b	6.79(0.04) b	309.9(23.0) b
	Floor mgmt (F)					
	Mulch	17.9(1.9) a	11.8(0.7) a	627.5(45.3) a	6.86(0.03) a	239.9(15.9) a
	Clean	77.1(7.5) b	6.4(0.5) b	373.2(23.0) b	6.95(0.04) b	410.4(18.9) b
	Pr > F					
	Irrigation (I)	ns	ns	ns	ns	ns
	N-source (N)	ns	**	****	***	**
	Floor mgmt (F)	****	****	****	*	****

[x] Numbers in parentheses following mean are standard error of the mean, followed by different lowercase letters within columns indicating differences of least squares means using the Tukey–Kramer adjustment, at $p < 0.05$.*, **, ***, **** and ns indicate that means are significantly different at $p \leq 0.05, 0.01, 0.001, 0.0001$ or are not significantly different, respectively.

In 2013, none of the management factors affected soil NH_4^+–N concentrations (Table 5). In 2014, soil NH_4^+–N concentrations in Compost and Mulch plots were significantly higher than Urea and Clean plots, respectively. Most of the mineral N in soil was in the form of NO_3^-–N (83% averaged across all treatments over two years); soil NH_4^+–N concentrations were low (mostly less than 10 $mg\ kg^{-1}$). On average soil NH_4^+–N concentrations increased by 35% from 2013 to 2014.

In both years, soil SEOC concentrations were not affected by the type of irrigation imposed (Table 5). However, the soil SEOC concentrations in Compost and Mulch plots were about 1.6 ($p < 0.0001$) and 1.4 times higher

($p < 0.05$) than Urea and Clean plots, respectively. On average soil SEOC concentrations increased by 12% from 2013 to 2014.

In 2013, only the type of N-source used influenced the soil pH; the pH in Compost plots was higher than Urea plots (Table 5). In 2014 floor management affected pH in addition to N-source; the pH in Clean plots was higher than Mulch plots. The soil became more acidic between 2013 and 2014; soil pH was 7.4 in 2011 before treatment initiation and pH averaged 7.2 and 6.9, in 2013 and 2014 respectively, across all treatments.

In both years, soil EC was not affected by the type of irrigation used. However, the soil EC was significantly affected by N-source ($p < 0.01$) and floor management ($p < 0.0001$). EC in Compost > Urea and Clean > Mulch plots. On average soil EC increased by 47% from 2013 to 2014, consistent with the increment of measured ionic nutrients (NO_3^-–N and NH_4^+–N concentrations).

4. DISCUSSION

4.1. Seasonal N_2O emissions: thaw and the pre-growing season

A characteristic pattern of N_2O emission during two principle periods of higher emissions was observed in both years for all treatments: during freeze–thaw cycles in the PreGS and; during fertigation in the GS. On average, a considerable portion (37% in 2013 and 61% in 2014) of the annual cumulative N_2O emission occurred during the PreGS particularly during the thaw period. Prior studies in perennial cropping systems expected minimal N_2O emission in winter and hence focused their N_2O monitoring only during the GS (Sanchez-Martin et al., 2008, Sanchez-Martin et al., 2010, Smart et al., 2011 and Suddick et al., 2011). However, similar to our observation, significant winter N_2O emissions were observed for other cropping systems in climates that experience spring ground thaw in western Canada (Nyborg et al., 1997, Lemke et al., 1999 and Izaurralde et al., 2004) and elsewhere (Müller et al., 2002, Holst et al., 2008 and Tatti et al., 2014).

In a recent review, Risk et al. (2013) elucidated two mechanisms for enhanced N_2O emissions during the thaw period: (1) the physical release of N_2O produced throughout the winter by the removal of ice blocking soil pores and; (2) the emission of newly produced N_2O at the onset of thaw due to increased biological activity and changes in physical and chemical soil conditions. We observed over a month of heighted N_2O emission during the

PreGS in both years, suggesting that the physical release of N_2O produced during the winter is likely minor and so the contribution of this mechanism to the total cumulative PreGS N_2O emissions in this system is likely minor. The occurrence of enhanced denitrification by the second mechanism during freeze–thaw depends mainly on the prevalence of anaerobic conditions (Groffman et al., 2009). During freeze–thaw events, it is likely that WFPS exceeded the WFPS threshold needed to create anaerobic conditions resulting in N_2O emission spikes. WFPS in our study appears low (<45%) during freeze–thaw cycles but the actual WFPS was likely higher in the top 0–5 cm of the soil, as the majority of the 30 cm long TDR probe was still under frozen ice at deeper depth, and any frozen water would not be measured.

Spring emissions may also depend on the bacterial activity, and availability of substrate. Increased abundance of cold-adapted nitrifier and denitrifier communities in late winter, especially in soils receiving organic amendments, has also been suggested to cause thaw period N_2O pulses (Tatti et al., 2014). Cell lysis in plant roots and soil microbes on freezing (Herrmann and Witter, 2002) may also provide fresh substrates for surviving microbes. This mechanism may enhance the level of denitrification by both fueling denitrifiers and depleting O_2 (Mørkved et al., 2006).

Some studies link greater spring N_2O emissions following lower temperatures during early winter and attribute this to more cell lysis (Neilsen et al., 2001 and Koponen and Martikainen, 2004). In our study, emissions in the pre-GS of 2014 were 83% higher than 2013, following a colder winter than 2013. The lower air temperatures measured prior to thaw in 2014 (lowest recorded winter temperature = −12.7 °C) than in 2013 (lowest recorded winter temperature = –8 °C) may indicate greater cell lysis in early winter. The 2014 thaw period was also longer in duration and included multiple freeze thaw cycles, which may have promoted additional cell lysis during the thaw period.

4.2. Seasonal N_2O emissions: growing season

Total growing season emissions were reduced between 2013 and 2014 despite a substantial increase of the mean growing season soil NO_3^-–N and NH_4^+–N concentrations. The difference in the growing season N_2O emissions was likely related to the dynamics of soil mineral N and moisture. During the GS of 2013, the highest daily fluxes (19–29 N $ha^{-1}day^{-1}$) occurred following intense rainfall that coincided with the end of fertigation when soil nitrate was at a maximum, creating potential for high water saturation, low oxygen diffusion rate, conditions suited to denitrification. There was no such concurrent occurrences of intense rain and enhanced soil mineral N in the growing season of 2014. The highest 2014 daily fluxes recorded around mid-fertigation period during the growing season and the values (7–12 N $ha^{-1}day^{-1}$) were much smaller than 2013.

The distribution of nitrogen within plots may have contributed to the N_2O emission spike seen under drip, but not micro-sprinkler, irrigation in July and August 2013. The wetter zone under drippers has higher root densities (Neilsen et al., 1997) and higher N concentrations during fertigation. The wetter zone expands over the fertigation season as more water is applied thereby expanding the flux zone as a result of the transport of N and degradation of organic matter away from the drippers (Sanchez-Martin et al., 2008 and Sanchez-Martin et al., 2010). When fertigation ceases, but irrigation continues, the continued addition of fresh water via the dripper may transport N both laterally and downwards (Hanson et al., 2006) away from the zone of root concentration under the dripper and towards the plot margins where the N is less likely to be taken up. Consequently, in Drip plots, the N component of lateral transport may be available near the soil surface for denitrification after fertigation ceases. In contrast, the flux zone under micro-sprinkler irrigation is more uniform during fertigation because nitrate and roots are more evenly distributed. Residual N in soil at the end of microsprinkler fertigation is pushed downwards into the soil by additional irrigation water. Most of N_2O generated at deeper depths from residual N is more likely to be consumed during diffusive transport to the soil surface (Kellman and Kavanaugh, 2008).

The intense rainfall events in late June 2013, when soil nitrate was maximum, may also have affected the location of nitrous oxide fluxes and total growing season emissions. It was noted the rainfall intensity exceeded the infiltration capacity and was sufficient to cause runoff from the rows towards the alleys, following the slightly mounded shape of the soils in the rows. This may have facilitated greater lateral movement of nitrate away from the drippers in that year, moving N earlier into soil with lower N uptake by roots, and increasing the zone of N_2O emissions. In micro-sprinkler, lateral runoff would not distribute N to previously N free areas, and did not result in increased emissions. This emission spike in July and August 2013 in drip irrigation was consistent with increases in soil NO_3^-–N availability (determined biweekly using ion-exchange resins) under drip irrigation later in the season, while no similar spikes in soil NO_3^-–N availability were observed under micro-sprinkler (Hannam et al., 2016). No such similar N_2O spikes were noted in the irrigation season of 2014. There was only one intense rainfall during fertigation in that year, occurring earlier in the season (June 13) well before the soil nitrate was expected to reach its maximum (End of June). This may not have been sufficient to cause lateral migration of nitrogen away from the dripper.

4.3. Interannual variations

Cumulative N_2O emissions in 2014 were over 14% higher than in 2013. The analysis of seasonal effects indicates that the majority of the interannual variation relates to the thaw period, during which emissions increased between years by 83%. The growing season emissions were reduced by 40% between 2013 and 2014. The variability in the total amounts of natural precipitation between 2013 and 2014 had little direct effect on the interannual variation on N_2O emission as irrigation additions resulted in WFPS being similar during the GS of both years. The intense rain events that occurred when soil nitrate was at its maximum around the end of fertigation in 2013 likely caused the higher GS season emission in that year.

We also noted interannual variations in nitrogen availability, soil temperature and soil pH. Soil NO_3^-–N and NH_4^+–N concentrations in the growing season both increased substantially from 2013 to 2014. The increase in the soil mineral-N may have been driven by two factors: accumulation of soil mineral-N when fertilizer application exceeded plant demand and/or changes in the total denitrification activity due to decreasing pH. The majority of the applied N (either urea or compost) in our system would first undergo ammonification, which involves enzymatic conversion of organic forms of N to NH_4^+. Some of the NH_4^+ would have been directly absorbed by the plant while the majority of the remainder would be oxidized to NO_2^- and then to NO_3^-, the most suitable nitrogen form for plant growth (Miller and Cramer, 2005 and Rothstein and Cregg, 2005). Ammonium oxidation is an acidifying process ($NH_4^+ + 1.5O_2 \rightarrow NO_2^- + H_2O + 2H^+$) (Van Dongen et al., 2001); which is consistent with our observation of decreasing soil pH over the years. This decrease in pH thus has a potential to cause accumulation of soil NO_3^- as decreased pH would lead to reduced complete denitrification to N_2. Decreasing soil pH leads to increase of the $N_2O/(N_2O + N_2)$ ratio with no changes in N_2O production (Čuhel et al., 2010); which means a reduction in the portion of soil NO_3^- that would be converted to N_2. Thus decreasing soil pH may promote the increase in soil NO_3^- which may eventually lead to higher N_2O amounts on an annual basis by increasing N availability in thaw periods and intense rainfall events. Our results match others (Schellenberg et al., 2012, Asgedom et al., 2013 and Xie et al., 2014) who observed a positive relationship between soil mineral-N and N_2O emission on an annual basis.

Overall, the 14% interannual variability in the amounts of N_2O emission in our study is within the range of Verhoeven and Six (2014) who observed 2–58% variability of N_2O in a 2-year study of a wine grape system amended with N (organic and inorganic) and biochar in California. The Merlot vineyard in our experiment was only 3-years old; roots and vines may have still been establishing. We noted a reduction in the variability of N_2O emissions between treatments between the second and third year of the vineyard history. The pH, soil NO_3^-–N, NH_4^+–N, SEOC and EC changing over the years may imply that the soil is not at steady state and hence may indicate the need for longer term N_2O monitoring.

4.4. Soil and water management effects on N_2O emissions

Surface application of bark mulch decreased 2-year mean $\sum N_2O$ emission by 28%. To the best of our knowledge there is only one published study, Livesley et al. (2010), that investigated the effect of bark mulch on N_2O emission using direct manual chamber measurements at the field scale. Livesley et al. (2010) observed 50% lower $\sum N_2O$ emissions from wood-chip mulched irrigated garden than irrigated lawn in Australia, but did not compare garden with and without wood chip mulch.

A combination of factors may be responsible for the lower N_2O emission in bark Mulch plots compared to Clean plots. Under the bark mulch, we found significantly higher organic carbon similar to other studies (Livesley et al., 2010 and Neilsen et al., 2014). This study also found higher WFPS, and lower soil nitrate concentration, which complements the lower available nitrogen measured by soil exchange resins found byHannam et al. (2016). The higher WFPS and added carbon substrate may have created favourable anaerobic conditions for enhanced complete denitrification to N_2, resulting in lower $\sum N_2O$ emissions under the bark mulch. Microbial immobilization of mineral N increases in the presence of an available labile carbon source (Homyak et al., 2008); hence, the reduced mineral-N availability under mulch could have been a result of microbial immobilization (Homyak et al., 2008). Lower levels of mineral N might have lowered the $\sum N_2O$ emissions under mulch even further.

Similarly, the lower $\sum N_2O$ emission in Compost than Urea in 2013 may be partly attributed to enhanced complete denitrification because of the higher soil carbon substrate in Compost plots. Our results were similar to those reported by Alluvione et al. (2010) who found lower N_2O emissions for municipal solid waste compost than mineral fertilizer in a silt loam soil under a temperate subcontinental climate. In both the 2014 and the combined 2-year data, we did not see differences in annual $\sum N_2O$ emissions between compost and urea plots. The increase of EC in soil under compost by 41% from 2013 to 2014 might have partially limited the potential for of the higher soil carbon under compost to enhance complete denitrification to N_2, as increased EC has been reported to reduce the activities of N_2O reductase (Ruiz-Romero et al., 2009).

Micro-irrigation resulted in differences in 2-year mean $\sum N_2O$ emission only during the growing season; $\sum N_2O$ emission under micro-sprinkler was 29% lower than drip irrigation. This difference was likely caused by increased frequency of high WFPS in Drip over Micro-sprinkler plots. Bateman and Baggs (2005) observed higher emissions with WFPS > 60%. During the growing season, the Drip plots exceeded 60% WFPS 21% and 30% of the time in 2013 and 2014, respectively, compared to Micro-sprinkler plots where WFPS was recorded above 60% only 7% and 4% of the time in 2013 and 2014, respectively. These water contents are measured at 30 cm lateral distance from the location of the dripper and thus water contents directly under the dripper can be expected to be higher. Those under micro-sprinkler irrigation are representative of the whole plot. The effect of irrigation types on subplot-scale N_2O emissions and on microbial pathways for the production of N_2O in our study site is being investigated in parallel studies. Early results of the microbial study indicate that drip irrigation reduces the abundance of N_2O reducers (abundance of nosZ genes) compared to irrigation with microsprinklers (Voegel, personal communication). The higher WFPS observed under drip irrigation might have limited the activity of the enzyme nitrous oxide reductase (nosZ) to convert N_2O to N_2 gas, thus resulting in higher N_2O emissions.

Our observation of higher N_2O emission with drip irrigation was similar to the findings ofSmart et al. (2011), who observed approximately 60% lower $\sum N_2O$ emission under fanjet sprinkler irrigation than drip irrigation in a Mediterranean climate during the growing season. Their vineyard was irrigated twice a week and received 34 kg N ha^{-1} year^{-1}. The N_2O mitigation potential of micro-sprinkler irrigation was similar in our studies. However,Smart et al.'s (2011) measurements lasted for only a month in the fall and another month in spring during fertigation and the longer-term N_2O mitigation potential of micro-sprinkler in their system is unknown. Our findings of major pre-growing $\sum N_2O$ emissions, and the significant effect of micro-irrigation in the 2-year mean growing season $\sum N_2O$ emission (but not annual) indicates the importance of long-term (≥2 year) continuous year round monitoring across seasons.

4.5. Yield-scaled N_2O emission and N_2O emission factor

Studies that only consider area-scaled N_2O emission cannot be used to recommend management practices that are suitable for mitigation of N_2O emission without compromising agricultural productivity. In our study, both area-scaled N_2O emission and yield-scaled N_2O emissions were significantly decreased by surface application of bark mulch. Surface application of bark mulch reduced 2-year mean yield-scaled N_2O emissions by 23%. Thus, bark mulch can be used to mitigate N_2O emission without compromising productivity. To the best of our knowledge, this study represents the first report of this metric (i.e., yield-scaled N_2O emissions) being generated for a Merlot grape vineyard. Grapes used for wine production have a more complex relationship between production mass and eventual total wine value (i.e., increased yield may not be as important as optimal fruit quality parameters); further exploration of yield-scaled or value-scaled emissions may be needed.

From our sandy loam soil across treatments over 2 years, the mean fraction of applied N lost as N_2O (N_2O emission factor), uncorrected for background emission, ranged from 2.53 to 3.14%. We could not calculate the true emission factors (i.e., corrected for the background emission) because there was no control plot as vines with no treatment (such as irrigation) would not survive in our semi-arid climate. Thus true emission factors would likely have been slightly lower than these values. Emission factors in our study were at least twice greater than the 1% default direct emission factor of the Intergovernmental Panel on Climate Change (IPCC) methodology for calculating N_2O emission rates from agriculture irrespective of fertilizer and soil type (IPCC, 2006). The 1% IPCC default direct emission factor may be overly simplistic for micro-irrigated systems receiving fertigation and organic amendments. In perennial cropping systems water and nitrogen are usually applied in smaller area of soil close to vine/tree compared annual crops. This practice increases nitrogen and water use efficiency. However, this also leads to increased localized concentration of nitrogen especially in drip irrigated system as compared to other applications such as broadcast and furrow irrigation (Smart et al., 2011). Hence, there is a higher potential for enhanced denitrification and the 1% IPCC emission factor used to calculate N_2O emission from agriculture would likely underestimate estimation from perennial cropping systems.

5. CONCLUSION

This study demonstrated the important contribution of the pregrowing season, particularly the thaw period, to the annual N_2O emission in vineyards under a semi-arid climate with cold winters. Disregarding this period could lead to either serious underestimations of annual N_2O emissions or/and bias in treatment comparisons because effects could be variable across seasons in a year. More work is needed in climates with and without winter freezing temperatures to assess annual flux changes.

Two year results indicate that micro-irrigation type affected growing season emissions. N-sources used in our study did not affect N_2O emissions significantly ($p > 0.05$). Surface application of bark mulch in vineyards can reduce both annual area-scaled and yield-scaled N_2O emissions. The mechanism by which bark mulch reduces N_2O emissions is not clear and further investigation using isotopic and/or molecular techniques is warranted. Further work is also required to determine the life-cycle consequences of carbon emissions from the bark mulch on overall greenhouse gases emissions. Published studies that investigated interannual variability in N_2O emissions from woody perennial systems are rare and the current study is one of the first works to give insights in interannual variability for these systems. Our study has one of the longest monitoring durations for experiments examining the effects of micro-irrigation, N-source and mulching in perennial systems. However, two years may still be a short time to fully understand long-term changes as C and N status of the soil changed over time, particularly in response to bark mulch decomposition.

ACKNOWLEDGEMENTS

This work was funded by the Agricultural Greenhouse Gases Program of Agriculture and Agri-Food Canada (AGGP-AAFC), the Canada Foundation for Innovation (CFI), and work-study program of University of British Columbia (UBCWS). The authors gratefully acknowledge the contributions of: Marina Molodovskaya,Valerie Ward, Scott Fazackerley and Markandu Anputhas at UBC; Shawn Kuchta, Istvan Losso, Bill Rabie and student field crews from AAFC; and several university co-operative education students (Brittany Derrick, Russell Kirchner, Tom Zochowski, Chris Perra, Mirage Leung) and UBCWS students (Graham Knibbs, Jeff Kerkovius, Stephen Kimanzi, Dan Lang, Leo Kerrigan, Danielle Homer, Jeffrey van Santen, Caroline Hedge) and Robert Roskoden (German Government DAAD internship student).

REFERENCES

1. ReferencesAlluvione, F., Bertora, C., Zavattaro, L., Grignani, C., 2010. Nitrous oxide and carbondioxide emissions following green manure and compost fertilization in corn.Soil Sci. Soc. Am. J. 74, 384–395.

2. Asgedom, H., Tenuta, M., Flaten, D.N., Gao, X., Kebreab, E., 2013. Nitrous oxideemissions from a clay soil receiving granular urea formulations and dairymanure. Agron. J. 105, 1–13.

3. BCGA, 2014. British Columbia Grapegrowers Association. B.C. Wine Grape AcreageReport [Online] (accessed 12.03.16). Available: http://grapegrowers.bc.ca/pdf/2014AcreageSurveyReport.pdf.

4. Bateman, E.J., Baggs, E.M., 2005. Contributions of nitrification and denitrification toN2O emissions from soils at different water-filled pore space. Biol. Fertil. Soils41, 379–388.

5. Box, G.E.P., Cox, D.R., 1964. An analysis of transformations. J. Roy. Statist. Soc. Ser. B26, 211–252.

6. Brown, C., 2012. Manure Sulphur – Digging Deeper [Online] (accessed 12.03.16).Available: http://www.regionalscia.org/Croptalknov12.pdf.

7. Carlisle, E., Smart, D., Williams, L.E., Summers, M., 2010. California VineyardGreenhouse Gas emissions: Assessment of the Available Literature andDetermination of Research Needs. California sustainable wine growing alliancepublication, San Francisco, CA [Online] (accessed 07.7.14). Available: http://www.sustainablewinegrowing.org/docs/GHGreport.pdf.

8. Chantigny, M.H., Angers, D.A., Kaiser, K., Kalbitz, K., 2008. Extraction andcharacterization of dissolved organic matter. In: Carter, M.R., Gregorich, E.G.(Eds.), Soil Sampling and Methods of Analysis Canadian Society of Soil Science.CRC Press, Boca Raton, Florida, pp. 617––635.

9. Chantigny, M.H., Rochette, M.H.P., Angers, D.A., Bittman, S., Buckley, K., Massé, D.,Belanger, G., Eriksen-Hamel, N., Gasser, M.O., 2010. Soil nitrous oxideemissions following band-incorporation of fertilizer nitrogen and swinemanure. J. Environ. Qual. 39, 1545–1553.

10. Cochran, V.L., Sparrow, E.B., Schlentner, S.F., Knight, C.W., 1997. Long-term tillageand crop residue management in the subarctic: fluxes of methane and nitrousoxide. Can. J. Soil Sci. 77, 565–570.ˇ

11. Cuhel, J., Simek, M., Laughlin, R.J., Bru, D., Che'neby, D., Watson, C.J., Philippot, L.,2010. Insights into the effect of soil pH on N2O and N2emissions and denitrifiercommunity size and activity. Appl. Environ. Microbiol. 76, 1870–1878.

12. Environment Canada, 2014a. Canadian climate normals or averages, forSummerland and Penticton, BC [Online] (accessed 13.02.14) Available: http://climate.weather.gc.ca/climate normals/results 1981 2010 e.html.

13. Environment Canada, 2014b. Canadian climate data on-line customized search.Canada's National Climate Archive, for

Summerland and Penticton, BC[Online](accessed 13.02.14) Available:
http://www.climate.weatheroffice.gc.ca/advanceSearch/searchHistoricData e.html.

14. FAO, 2013. Food and Agriculture Organization of the United Nations StatisticsDivision [Online](accessed 15.11.15) Available: http://faostat3.fao.org/download/Q/QC/E.

15. Forge, T.A., Hogue, E.J., Neilsen, G., Neilsen, D., 2008. Organic mulches alternematode communities, root growth and fluxes of phosphorus in the root zoneof apple. Appl. Soil Ecol. 39, 15–22.

16. Forge, T.A., Temple, W., Bomke, A.A., 2013. Using compost as mulch for high bushblueberry. Acta Horti. (ISHS) 1001, 369–376.

17. Gagnon, B., Ziadi, N., Rochette, P., Chantigny, M.H., Angers, D.A., 2011. Fertilizersource influenced nitrous oxide emissions from a clay soil under corn. Soil Sci.Soc. Am. J. 75, 595–604.

18. Gale, E.S., Sullivan, D.M., Cogger, C.G., Bary, A.I., Hemphill, D.D., Myhre, E.A., 2006.Estimating plant-available nitrogen release from manures, composts, andspecialty products. J. Environ. Qual. 35, 2321–2332.

19. Garland, G.M., Suddicka, E., Burgerb, M., Horwathb, W.R., Six, J., 2011. Direct N2Oemissions following transition from conventional till to no-till in a covercropped Mediterranean vineyard (Vitis vinifera). Agric. Ecosyst. Environ. 144,423–428.

20. Groffman, P.M., Butterbach-Bahl, K., Fulweiler, R.W., Gold, A.J., Morse, J.L., Stander,E.K., Tague, C., Tonitto, C., Vidon, P., 2009. Challenges to incorporating spatiallyand temporally explicit phenomena (hotspots and hot moments).Biogeochemistry 93, 49–77.

21. Halvorson, A.D., Del Grosso, S.J., Alluvione, F., 2010. Nitrogen source effects onnitrous oxide emissions from irrigated no-till corn. J. Environ. Qual. 39,1554–1562.

22. Hannam, K.D., Kehila, D., Millard, P., Midwood, A.J., Neilsen, D., Neilsen, G.H., Forge,T.A., Nichol, C., Jones, M.D., 2015. Bicarbonates in irrigation water contribute tocarbonate formation and CO2production in orchard soils under drip irrigation.Geoderma

23. Hannam, K.D., Neilsen, G.H., Forge, T., Neilsen, D., Losso, I., Jones, M.D., Nichol, C.,Fentabil, M.M., 2016. Irrigation practices, nitrogen amendments and mulchesaffect nutrient dynamics in a young Merlot (Vitis vinifera L.) vineyard. Can. J.Soil Sci.,

24. Hanson, B.R.,ˇSim˚unek, J., Hopmans, J.W., 2006. Evaluation ofurea–ammonium–nitrate fertigation with drip irrigation using numericalmodeling. Agric. Water Manage. 86, 102–113.

25. Herrmann, A., Witter, E., 2002. Sources of C and N contributing to the flush inmineralization upon freeze–thaw cycles in soils. Soil Biol. Biochem. 34,1495–1505.

26. Holst, J., Liu, C., Yao, Z., Brüggemann, N., Zheng, X., Giese, M., Butterbach-Bahl, K.,2008. Fluxes of nitrous oxide, methane and carbon dioxide duringfreezing-thawing cycles in an Inner Mongolian steppe. Plant Soil 308, 105–117.

27. Homyak, P.M., Yanai, R.D., Burns, D.A., Briggs, R.D., Germain, R.H., 2008. Nitrogenimmobilization by wood-chip application: Protecting water quality in anorthern hardwood forest. For. Ecol. Manage. 255, 2589–2601.

28. Hutchinson, G.L., Livingston, G.P., 1993. Use of chamber systems to measure tracegas fluxes. In: Rolston, D.E. (Ed.), Agro-Ecosystem Effects on Trace Gases and

2
CHAPTER

Limits to Agricultural Growth in the Sistan Closed Inland Delta, Iran

Eelco van Beek[1,2], Babak Bozorgy [1], Zoltán Vekerdy [2] and Karen Meijer[3]

[1]*Delft Hydraulics, P.O. Box 177, 2600 MH Delft, The Netherlands*

[2]*University of Twente, P.O. Box 217, 7500 AE Enschede, The Netherlands*

[3]*Water Research Institute, Ministry of Energy, P.O. Box 16765-313, Tehran, Iran*

[4]*Institute for Geo-Information Science and Earth Observation, P.O. Box 6, 7500 AA Enschede, The Netherlands*

ABSTRACT

The Sistan Delta in Iran is located at the end of a closed basin with nearly 100% of the supply coming from Afghanistan. This supply is supporting irrigated agriculture in the area and is the source for the lake system around the delta. These Hamoun lakes are ecological very valuable wetlands; a number of them are registered as Ramsar sites. The Iranian government tries to improve the living conditions of the people in the area, among others by providing infrastructure for irrigated agriculture. Further development of the irrigated area will mean less water for the Hamouns with resulting lower average water coverage of the lakes. This will not only endanger the ecosystem that the Hamouns support but also the livelihoods of the people that depend on the goods and services that the lakes provide. This paper describes a study that has been carried out to support decision making on potential development schemes in the delta. The non-availability of data from Afghanistan requires the development of various tools and the use of remote sensing techniques to enable to make estimates for the river flow that Iran can expect from Afghanistan. An IWRM approach has been used for the balancing of interests involved. Some preliminary conclusions are described.

Keywords: Integrated water resources management Irrigation Remote sensing Wetlands

INTRODUCTION

Wetlands in semi-arid and arid regions are usually fed by rivers, which bring the water from remote areas, e.g. the Mesopotamian Marshes or the Okavango Delta. People living in these arid regions heavily depend on the discharge of the rivers and the water stored in the wetlands: their agriculture is irrigation based since there is no sufficient rainfall to support production. This is the case of the Sistan basin too.

THE STUDY AREA

The Sistan inland delta is a densely populated enclave in the south-eastern part of Iran at the lower end of the Hirmand River (Fig. 1). In Afghanistan this river is called the Helmand. The entire contributing basin is about 200,000 km^2 and is largely located in Afghanistan. The Iranian part, the delta plain (ca. 2,500 km^2) and part of the surrounding wetlands system (ca. 5,000 km^2), covers less than 5% of the total basin area. The river system is fed by rains and snowmelt in the Hindu-Kush Mountains and discharges into an inland depression which, when sufficient water is available, forms the Hamoun Lakes. These lakes are one of the main and most valuable aquatic ecosystems in Iran. About half of the area is registered in the Ramsar and UNESCO Biosphere Reserve Conventions. A unique feature of the lakes is that they are fresh or brackish, despite that they are seemingly at the end of a closed basin. Actually, they are not the end. During periods of very high flows the lakes spill via the Shile River into the Goud-e-Zereh. This 'flushing' happens on average each 8–11 years. The Goud-e-Zereh is the real terminal lake of the basin and is very saline.

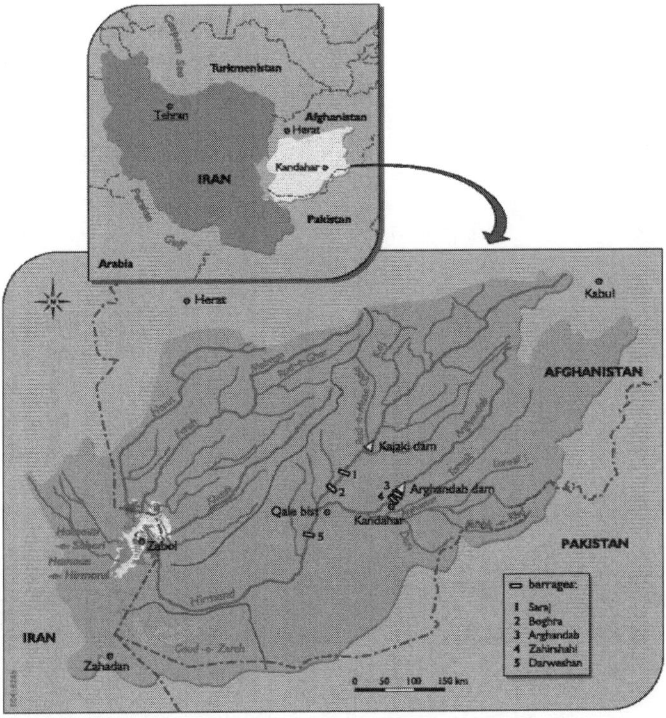

Figure 1. Contributing river basins Sistan inland Delta

The Sistan inland delta has a population of some 400,000 people on the Iranian side, whilst only a few ten thousands live on the Afghan side. The economy is strongly dependent on agriculture and the goods and services provided by the wetlands. The irrigation system of about 120,000 ha has recently been rehabilitated. Three reservoirs (Chahnimeh) have been constructed for public water supply with a fourth reservoir under preparation. Details on the Sistan delta and the Hamoun lakes are given in Fig. 2.

The Sistan Delta has a very hot and dry climate. In summer the temperature exceeds 50°C. Rainfall is about 60 mm/year and occurs only in autumn and winter. The open water evaporation is very high and is estimated at 3200 mm/year. Some local sources even claim amounts up to 5000 mm/year. Strong winds are an important reason for this high evaporation. In summer, a phenomenon occurs that is called the '120 days wind' with prevailing wind speeds of nearly 10 m/s (Arasteh et al. 2007).

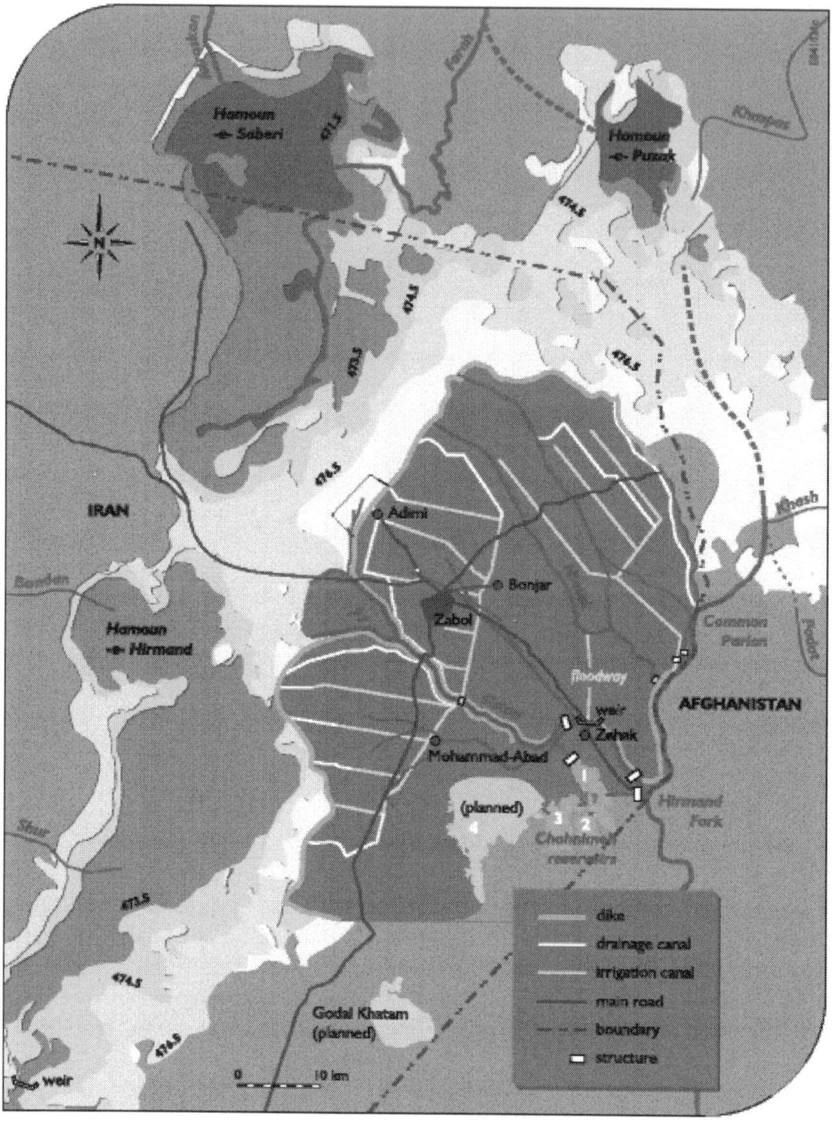

Figure 2. Overview map of the Sistan Inland Delta

Sustainability issues

The socioeconomic and ecological sustainability of the area is a major issue. There are several reasons for this. First of all the potential of the area for socioeconomic developments is limited. Further growth of the traditional activities such as agriculture and the use of the wetlands (fishing, reed cutting and bird catching) is severely constrained by a lack of reliable water supply and the limited carrying capacity of the natural system. The high increase in population (natural and through migration from Afghanistan) has caused a very high pressure on the system and over-use and mismanagement have resulted in a collapse of the natural system. Intensified cattle grazing and the introduction of exotic herbivorous fish have destroyed the regeneration capacity of the reeds after dry years. This is aggravated by the highly dynamic river flow with strong yearly fluctuations as well as with wet and dry periods lasting several years (Fig. 3). The wet period between 1989 and 1993 caused considerable flooding damage in the area. The dry period between 2000 and 2005 resulted in the Hamouns falling completely dry, turning them into a desert and wiping out the water related ecosystem in the area. Mismanagement of the system (in particular the introduction of grass carp that destroyed the reed fields) has worsened the situation and has made it more difficult for the system to recover after the dry period.

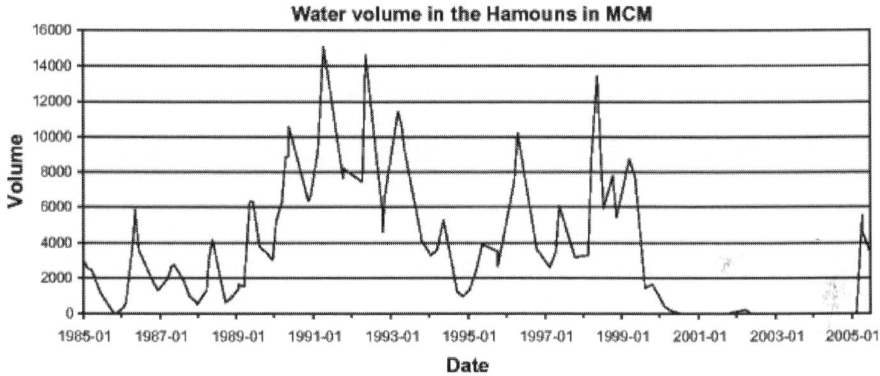

Figure 3. Water volume in the Hamouns in MCM in period 1985–2005

Another threat to the sustainability of the area is the potential developments in Afghanistan. Nearly all water for the Sistan delta originates from Afghanistan. Many plans exist in Afghanistan to develop the water resources upstream (reservoirs, irrigation systems, drinking water withdrawals, etc.). No reliable information exists on the status of these developments but it is highly likely that some if not all of these developments will take place sooner or later. The existing agreement between Afghanistan and Iran on the Hirmand is rather limited in scope and only guarantees Iran a minimal amount of water sufficient for drinking water purposes.

Objective of the study

The Iranian government is trying hard to improve the living conditions of the people in the area, in particular by providing employment and improving the health situation of the people. The available water resources is a constraint for this development, in particular for the planned extension of irrigated agriculture. The objective of the study was to quantify the supply and potential demand, design alternative development strategies while protecting the ecological values of the wetlands. This paper focuses on the potential development of agriculture in the area in the mirror of water availability and balancing of interests needed to take decisions in this respect. A main part of the study consisted of the development of tools to predict the water supply and to quantify the impacts of possible developments on the Hamouns. The study was jointly carried out by a Netherlands Consortium headed by Delft Hydraulics and the Water Research Institute (WRI) of Iran. Preliminary results are described in Delft Hydraulic and WRI (2006). The client of the study was the Sistan-Baluchestan Regional Water Authority (SBWRA).

DEVELOPING THE AREA – THE CHALLENGES

The main employment sectors in the Sistan delta are agriculture-fisheries (38%) and services (54%). The industrial sector is small (6%). Unemployment in the area is high: reliable statistics are lacking but indications point in the direction of about 50%. The very skewed population structure (about 50% of the population is below 20 years) will further degrade the situation. The Sistan Delta is a very political sensitive area bordering Afghanistan and Pakistan (Baluchestan) so the Iranian government tries to stabilize the area by investing in its socioeconomic development. An extensive system for drinking water supply has been constructed providing a near 100% coverage for urban and rural population. Possibilities to increase the industrial sector are constrained by the geographical location of the area in relation to potential markets of the products. Hence, it is not surprising that solutions are sought in agriculture. Already 120,000 ha irrigated agriculture have been developed and there are plans for another 125,000 ha. However, agricultural development is severely constrained by the harsh climatic conditions and the variable water supply. Moreover, the withdrawal of the water for irrigation will be at the expense of the Hamouns and their ecosystems. The variable river flow will require the development of storage reservoirs to support the irrigation. The existing Chahnimeh reservoirs (see Fig. 2) are only for drinking water but could be extended to serve irrigation as well. The total live storage capacity of all four Chahnimeh reservoirs together will be 950 MCM/year.

The water balance

The first question is if the available water supply will be sufficient to support these developments. Table 1shows a simplified average annual water balance for the delta for the present situation. The total available water is 5,935 MCM/year. The demand for the present 120,000 ha irrigated agriculture is 2,069 MCM/year. Due the variability of river flow and presently non-existent storage capacity for irrigation the actual supply is 1,169 MCM. If the irrigation area is increased this has to be done in combination with the development of surface water storage schemes. Groundwater storage is not possible because of the geo-hydrological situation.

Table 1. Average annual water balance Sistan Delta (in MCM/yr)

Inflow[a]		Outflow	
Hirmand	3,908	Agriculture (120,000 ha)[b]	1,161
Farah	1,512	Public water supply	34
Khash	455	Hamouns evaporation	4,378
Precipitation	60	Chahnimeh evaporation	124
		Outflow to Shile	238
Total	5,935	Total	5,935

[a] Derived from hydrological records of the period 1947–1971 at stations close to the inflow into the Hamoun system (resp. Char Burjak, Dilram and Farah) which are downstream of the major irrigation systems in Afghanistan

[b] Actual supply, the full demand of the 120,000 ha is much higher (2,069 MCM/yr); efficiency losses and drainage water are included as these do not return to the system (evaporate also); it includes also a small allocation to irrigation areas in Afghanistan

The Hamouns

The second question is what the effects of this additional agricultural development will have on the Hamouns. From a water quality point of view measures are taken, which take care that the pollution (fertilizers, pesticides) resulting from agriculture will not reach the lakes. A major part of the drainage water is collected in special basins from where the water evaporates. Hence, the main impact will be a reduction in the evaporation of the lakes and a reduction of the outflow to the Shile. The Hamouns are at present by far the largest 'water user' of the system evaporating 4,378 MCM/year. Reducing the evaporation of the Hamouns is synonymous with reducing the average water cover area of the lakes and with that their potential to support the ecology depending on that water. In principle the relation between agricultural development and water cover is quite simple. An open water evaporation of 3000 mm/year means that each 100 MCM/year additional withdrawal for irrigation will reduce the average lake area by 33 km^2 which is about 2.5% of the present average lake area. Hence, a full supply of the present 120,000 ha irrigation area with 2,069 MCM/year could reduce the average lake area with about 50%. In reality this relation is much more complex since the system consists of 3 parts, which are functioning relatively separately with their own highly variable supply. Because of this, the water cover of the lakes is very dynamic too. In dry periods the lakes can fall completely dry as has been the case in the period 2000–2005, temporary creating a desert-like environment. When the lakes start filling after a rainy period, the ecology recovers. Reed starts growing again and birds and fish return. This process has been going on for ages and proves that the system can absorb to some extent these natural dynamics. The question is if it also can absorb a more continuous decrease of inflow. A reduced water cover of the Hamouns will not only hurt the ecological functioning of the lake. Also the livelihoods of the people that depend on the products and services of this system are endangered.

The Hamouns provide goods and services for all people in the Sistan Delta. Four groups of values were identified for these people: income and food, health, perception and experience, and independence and social structure. A survey conducted among the stakeholders revealed that all people in the area (close *and*further away from the Hamouns, rural *and* urban) give great importance to these values with all of them agreeing (100% score) on the health effects involved. The combination of strong winds and water cover provide a kind of natural air conditioning while the same strong winds with empty Hamouns result in severe sand storms in the area. The overall conclusion of the survey was that all people in the delta considered the wetlands as very important for their lives and, hence, not only just the farmers and fishermen who are directly dependent on the lakes for their income.

APPROACH AND TOOLS

IWRM – balancing the interests

From the above it has become clear that a major issue for the study was to balance the benefits from irrigated agriculture with the values provided by the Hamouns. Integrated water resources management (IWRM) addresses, among others, this balancing of interests and has been applied as the basic approach for the study. IWRM is a process which promotes the coordinated development and management of water, land and related resources, in order to maximize the resultant economic and social welfare in an equitable manner without compromising the sustainability of vital ecosystems (GWP 2000). The systems analysis approach of IWRM takes care of a structured analysis process as well as the quantification of the aspects that need to be balanced. For the quantification a RIBASIM application for the Sistan delta has been developed. RIBASIM is a generic computer package for river basin allocation issues (Delft Hydraulics 2005). The analysis process follows the framework of analysis as described in Loucks and Van Beek (2005).

Predicting the supply

A crucial and at the same time one of the most unknown variable in determining the sustainability of the system is the supply of water from the Hindu-Kush Mountains. Existing data is limited and recent data is not available at all in Iran. More accurate data is needed to evaluate the feasibility of the development plans in Iran as well as to predict the long term discharge (one-year predictions e.g. for the determination of cropping areas and patterns) as well as the short term discharge (days till weeks for operational purposes, e.g. for flood warning and the operation of intakes). A complicating factor in predicting the supply is the unknown developments in the upstream areas. At present no data or other information is made available by Afghanistan to Iran. To cope with this situation a model system was developed that combined medium range weather forecasts with a grid-based hydrological model HBV (Bergström and Forsman 1973) modeled in PC-Raster and a river basin model for the river system in Afghanistan (based on RIBASIM).

Using remote sensing and GIS methods

Given the data scarcity involved in this case, both physically (limited monitoring in Afghanistan and Iran) as well as politically (the cooperation between Iran and Afghanistan on the Sistan wetlands just started recently) the use of earth observation data was very important in the study. The following three examples are given here.

Snow cover mapping using satellite images

The hydrological HBV module of the river runoff forecast model uses the weather forecasts from the European Centre for Medium Range Weather Forecast (ECMWF). The actual rain (and snow) fall will of course deviate from the estimate. Moreover, the grid size of this forecast is rather course.

An important variable that the HBV model calculates is the snow cover in Afghanistan, which determines to a large extend the water availability for the Sistan Delta. By using satellite images (Salomonson and Appel 2004), the calculated snow cover of the HBV model can be corrected, improving the accuracy of the forecast model.

For model calibration purposes, time series of snow cover maps from the 1990's were generated from NOAA AVHRR data. An automated procedure masked the cloudy areas from each selected archived image and then – using the altitude of the visible snow line in the cloud-free parts of the Hindu-Kush Mountains – created a snow cover map. This map was then used for improving the HBV model.

For operational river runoff forecast, 8-day snow cover products calculated from MODIS images (Hall et al.2002) are to be used, since these products are readily and freely available since the first years of this millennium via the World Wide Web.

Estimation of cropping area in Afghanistan

No reliable data is available on the actual cropped area in Afghanistan. For the identification of irrigated areas in arid regions such as Afghanistan an approximate solution is to map vegetation in the time of the year when natural vegetation is not green, leaving the irrigated areas to show up. This can be done by the normalized differential vegetation index (NDVI) which is calculated from the red and infrared bands of satellite images. Use is made of the MODIS sensor of the Terra and Aqua satellites that provides daily data and which are almost real-time accessible through internet (NASA 2005). For future conditions scenarios were developed by using information from FAO (1997) and Favre and Kamal (2004).

Inundation and vegetation cover of Hamouns

To increase our understanding of the hydrological and ecological functioning of the Hamoun systems, information was needed on the dynamics of the inundation and vegetation cover of the Hamouns. This was achieved by using Landsat and NOAA AVHRR data with some additional information retrieved from quick looks of IRS and Landsat. The results are described in Vekerdy et al. (2006).

ANALYSIS AND RESULTS

The analysis started with an inception phase in which – in interaction with the stakeholders – a joint problem statement and the development objectives for the area were formulated. The problem statement and objectives formed the base on which the criteria/indicators were determined that the stakeholders want to use in the evaluation of alternative development strategies. The indicators have a direct link with the objectives and include, among others:

- Reliability drinking of water supply (%)
- Supply/demand ratio of agriculture (%)
- Agricultural production (tons/year)
- Hamoun 'production' (fish, birds, reeds, in tons/year)
- Ecology of Hamouns
    - average area (ha)
    - flushing (frequency)
    - ecosystem condition (% of reference)
- Health
    - frequency of drying out of Hamoun-e-Saberi (causing sand storms)
- Income stakeholders
    - farmers – irrigation (frequency of years with < 50% reference income)
    - Hamoun users (same)

As reference for the income of the stakeholders and the ecosystem condition the situation in the decade 1970–1980 has been selected as this period was considered to be good, even though also in this period dry years occurred. Nature and people have adapted to the variability in water supply and they can survive 'normal' dry years. The indicators that use this reference express how much will be deviated from this.

The next step was to develop scenarios and strategies. A clear distinction was made between '*scenarios*' that represent external developments that cannot be influenced by the decision makers involved in the water resources in the Sistan area and '*strategies*' that include the interventions that these people can actually decide upon. The main scenario elements were population growth in the area and upstream developments in Afghanistan, more or less determining the boundary conditions with respect to demand and supply respectively. Strategy elements included supply oriented measures (such as additional reservoir capacity and the operation of the reservoirs), demand oriented measures (in particular related to agriculture), ecology oriented measures (legislation and enforcement, landscaping the Hamouns and Chahnimeh reservoirs to make them more suitable for ecology, etc.) and the increase/decrease of the area of irrigated agriculture.

The actual analysis process is still going on and is carried out in close cooperation with the stakeholders. Reference is made to the full documentation of the project that will be prepared by WRI and SBWRA. Below only the specific issue of agricultural growth will be described. The base case was the present situation: 120,000 ha irrigated area with only 3 Chahnimeh reservoirs to be used for public water supply only. The following agriculture-related strategies were developed:

Strategy 1the construction of Chahnimeh-4, to be used for public water supply only

Strategy 2same as strategy 1 but with allowed use of Chahnimeh-4 for irrigation also

Strategy 3an increase of the present irrigation area of 120,000 ha with an additional 125,000 ha, no use to be made of the Chahnimeh reservoirs for irrigation

Strategy 4same as strategy 3 but will Chahnimeh-4 and allowed use for irrigation

Strategy 5a decrease in irrigation area from the present 120,000 ha to 21.000 ha (in order to see the effects on the Hamouns)

All these strategies assume that the Hamouns are managed well and no over-exploitation of the natural resources takes place. As external scenario, the present situation in Afghanistan is assumed which includes the Kajaki and Arghandab reservoirs and about 240,000 ha irrigation. In this scenario, the average inflow from Afghanistan will be 5,795 MCM/year. To get some idea about the effects of upstream developments the base case is also evaluated for a scenario in which major upstream developments in Afghanistan will take place and the available water for Iran will reduce to about 3,257 MCM/year.

The impacts of these scenarios in terms of the defined indicators have been calculated by using the RIBASIM model as described above. A (partly reconstructed) time series of 55 years of inflows and climatological conditions (rainfall, evaporation) have been used to determine the average impacts. The average results are given in the score card of Table 2.

Table 2. Score card of alternative agricultural development strategies

Scorecard IWRM Sistan Inland Delta			Scenario 1						Scenario 2
Criteria	Unit	Target	Base case	1 Chahnimeh 4	2 Chahnimeh 4 for irrigation	3 Increase irrigated area	4 Incr. irr. with use Chah. 4	5 Decrease irrigated area	Base case
General information (on case)									
Irrigated area in Sistan	Ha	–	120,000	120,000	120,000	245,000	245,000	21,000	120,000
Average inflow from Afghanistan	MCM/yr	–	5875	5875	5875	5875	5875	5875	3250
Domestic water supply									
DMI raw water supply (Zabol + Zahedan)									
Months supply < demand	%	0	0	0	12.8	0	27.2	0	0
Economy – agriculture									
Supply/demand ratio	–	>0.80	0.63	0.53	0.79	0.42	0.45	0.87	0.31
Agricultural production	ktons/yr		401	340	580	570	634	111	348
Economy – Hamouns									
Fish	ktons/yr		20.7	20.7	20.8	20.8	20.8	20.6	4.8
Birds	ktons/yr		0.78	0.78	0.77	0.71	0.76	0.95	0.17
Reed	ktons/yr		299	299	296	295	296	303	133
Ecology – Hamouns									
Suitable conditions for ecology in Hamouns									
Average maximum area in fall (Oct–Jan)	ha		90,000	89,000	88,000	79,000	85,000	119,000	45,000
Flushing of Hamouns – Hirmand	T	<8 yr	12.5	12.5	12.5	16.7	25.0	7.1	16.7
Estimation of ecosystem condition	% of ref		90–100	90–100	90–100	73–87	73–87	90–100	0–20

Environment/Health									
Return per. in years Saberi <20% of max area	T	0	13	13	13	7	8	50	2
Income (return period of years with < 50% of reference income)									
Bird catchers	T		5.44	5.44	4.45	1.69	1.96	9.80	1.00
Fishermen	T		5.44	5.44	4.45	1.00	1.00	8.17	1.00
Reed harvesters	T		5.44	5.44	4.45	1.09	1.09	9.80	1.00
Pastoralists	T		5.44	5.44	4.45	1.69	1.96	4.90	1.00
Field cultivators near Hamoun	T		2.04	1.20	n/a	3.77	4.45	1.00	1.07
Animal husbanders on farm	T		1.20	1.20	n/a	n/a	n/a	1.00	1.11
Field cultivators far from Hamoun	T		1.20	1.20	n/a	n/a	n/a	1.00	1.11

As reference the situation in the 1970s has been selected as a period in which the conditions were considered to be good

T return period = 1/*f*

n/a not applicable; the situation did not occur in the 55 years of the simulation, and hence no return period could be assessed

Base case (present situation). Note the rather low supply/demand ratio for irrigation (0.63) which means that in many years there will be insufficient water available to cultivate the full present irrigation area.

Strategy 1 (Chahnimeh 4 constructed). The storage of water for public water supply will, as expected, deteriorate the conditions for agriculture somewhat. The conditions for the Hamouns hardly change.

Strategy 2 (Chahnimeh reservoirs also used for irrigation). The agricultural sector clearly benefits from this strategy. At the other hand the reliability of the public water supply system decreased. Optimization of the operation rules of the reservoirs (applying hedging rules) should be able to improve this.

Strategy 3 (increase of irrigation area without reservoir use). The agricultural production will increase somewhat compared to strategy 2 but the reliability of the supply drops to a level (0.42) that will be unacceptable for farmers. The ecological conditions of the Hamouns deteriorate clearly. The return period of a dry Hamoun-e-Saberi increases with corresponding dangers for sand storms. Also the return period of flushing of the Hamouns increases (once in 17 years now) which will result in an increase of the salt concentration of the Hamouns.

Strategy 4 (as strategy 3 but now with allowed use of the Chahnimeh reservoirs for irrigation). This strategy improves the performance of the agricultural sector somewhat but the reliability of the supply remains too low (0.45). This is caused by the fact that the reservoir capacity is limited. The reliability of the public water supply decreases dramatically but this can be improved by applying different operation rules.

Strategy 5 (decrease in irrigation area to 21,000 ha). As expected the agricultural production will drop but the reliability of the supply increases substantially which will be good for the remaining farmers. The ecological conditions improve considerably.

Besides these five strategies also the impacts have been calculated of a strong decrease of available water from Afghanistan (scenario 2). This is only done for the base case (present situation). From Table 2 it becomes clear that such scenario will have disastrous effects for the agricultural activities in the Sistan delta and for the Hamouns.

It is noted, that the effects of an increase of the demand in Afghanistan is much more than a similar kind of increase in Iran. This is due to the fact that a development in Iran will mainly influence the Hamoun-e-Hirmand, leaving the two other larger Hamouns (Hamoun-e-Puzak and Hamoun-e-Saberi) intact. Developments in Afghanistan will influence all three Hamouns and will result in a nearly complete collapse of the ecology of the Hamouns.

The above results should be considered with great care, in particular the conclusions with respect to the Hamoun-related effects. At the other hand, the calculated impacts seem reasonable and can be used to support the decision making process. Basically the following overall conclusions can be drawn:

- The use of water for irrigated agriculture is mainly restricted by the variability of the supply and not by the total supply.
- This makes sustainable irrigation of the present 120,000 ha in the Sistan Delta only possible if use can be made of the storage capacity of the Chahnimeh reservoirs for irrigation water also. Only in that case can a sufficient and reliable supply be guaranteed, which is needed for economically sound agricultural practices.
- Further increase of the irrigation area without the use of the Chahnimeh reservoirs does not make sense at all. Even if use can be made of the Chahnimeh reservoirs the reliability of the supply will be too low to sustain normal irrigated agriculture.
- There is a direct relation between the use of water for irrigated agriculture and the environmental conditions in the area. Water consumed by agriculture will not bc available anymore for the wetlands, resulting in a lower average water cover and corresponding impacts on ecology and health. Effluent water from the agricultural area poses a water quality related risk, but the discussion of this aspect is beyond the goals of this paper.
- Developments in the upstream areas can have disastrous impacts for the downstream areas. The Chahnimeh reservoirs will take care that the public water supply will not be endangered but the availability of water for irrigation will substantially decrease and the average area of the Hamouns will be much smaller. The frequency of a complete drying out of the Hamouns as happened in the period 2000–2005 will increase with all related problems for ecology and health (sandstorms).
- Developing the basin and the delta while at the same time protecting the Hamouns requires an intensive cooperation between Afghanistan and Iran.

DISCUSSION

The main aim of the study was to support the Iranian government in their decision making about the further development of the Sistan Delta. The geo-political situation requires that assumptions have to be made on the possible supply from Afghanistan both with respect to the natural physical aspects (rainfall-runoff) as well as the potential developments upstream (reservoirs, irrigation development). Further water consumption in the upstream areas endanger the downstream irrigation and ecological systems.

Remote sensing techniques have proven to provide valuable information for making such assumptions. If the boundary conditions from Afghanistan are considered to be fixed, the real issue for the decision makers in Iran is then to balance the benefits from irrigated agriculture (production, employment) against the loss in ecological and socioeconomic values of the reduced wetlands and related health hazards for the population. The analysis approach and tools developed in this study provide the quantitative information to support the balancing of interests. The ultimate choice is a political one.

REFERENCES

1. Arasteh PD, Tajrishy M, Mirlatifi M (2007) Statistical model of free water surface evaporation using the volume balance method in Chahnimeh reservoir, Sistan-Iran Pajouhesh & Sazandegi, http://www.runetwork.de/html/en/index.html?article_id=3814, Cited 15 June 2007
2. Bergström S, Forsman A (1973) Development of a conceptual deterministic rainfall-runoff model. Nord Hydrol 4:147–170
3. Delft Hydraulics (2005) RIBASIM version 6.33, River basin simulation model, Quick start guide
4. Delft Hydraulics WRI (2006) Integrated water resources management for the Sistan closed inland delta Iran. WRI, Iran
5. FAO (1997) Afghanistan Agricultural Strategy (Promotion of Agricultural Rehabilitation and Development Programmes for Afghanistan), TCP/AFG/4552
6. Favre R, Kamal GM (2004) Watershed Atlas of Afghanistan, First Edition, Working Document for Planners, Ministry of Irrigation, Water Resources and Environment, Afghanistan and FAO
7. GWP (2000) Integrated water resources management. Global Water Partnership Technical Committee (TEC), TEC Background Papers No. 4, Sweden
8. Hall DK, Riggs GA, Salomonson VV, DiGirolamo NE, Bayr KJ (2002) MODIS snow-cover products. Remote Sens Environ 83:88–89

9. Loucks DP, Van Beek E (2005) Water resources systems planning and management. UNESCO, Paris

10. NASA (2005) MODIS Rapid Response System. http://rapidfire.sci.gsfc.nasa.gov/subsets/?Afghanistan/, cited 16 June 2005

11. Salomonson VV, Appel I (2004) Estimating fractional snow cover from MODIS using the normalized difference snow index. Remote Sens Environ 89(3):351–360

12. Vekerdy Z, Dost RJJ, Reinink G, Partow H (2006) History of environmental change in the Sistan Basin based on satellite image analysis: 1976–2005. Report, United Nations Environmental Programme (UNEP) and ITC,http://postconflict.unep.ch/publications/sistan.pdf

3

CHAPTER

Improving Water-Efficient Irrigation: Prospects and Difficulties of Innovative Practices

Les Levidow[a,*], Daniele Zaccaria[b], Rodrigo Maia[c], Eduardo Vivas[c], Mladen Todorovic[d], Alessandra Scardigno[d]

[a]*Development Policy and Practice, Open University, UK, Milton Keynes MK7 6AA, UK*

[b]*University of California, Davis, CA, United States*

[c]*Universidade do Porto, Prac, a de Gomes Teixeira, Portugal*

[d]*CIHEAM—Mediterranean Agronomic Institute of Bari (IAMB), Bari, Italy*

ABSTRACT

Innovative irrigation practices can enhance water efficiency, gaining an economic advantage while also reducing environmental burdens. In some cases the necessary knowledge has been provided by extension services, helping farmers to adapt and implement viable solutions, thus gaining more benefits from irrigation technology. Often investment in technological improvements has incurred higher water prices, however, without gaining the full potential benefits through water efficiency. Farmers generally lack adequate means and incentives to know crops' water use, actual irrigation applications, crops' yield response to different water management practices, and thus current on-farm water-efficiency levels.

Those general difficulties are illustrated by our two case studies investigating options, stimuli and difficulties to improve water-efficient practices. The two areas have strong stimuli for improvement but lack a knowledge-exchange system to help farmers and resource managers identify scope for improvements. Partly for this reason, farmers' responsibility for efficient water management has been displaced to hypothetical prospects, e.g. extra supplies from reuse of treated wastewater or a long-term low water pricing. In both cases a displaced responsibility complements the default assumption that farmers' irrigation practices already have adequate water-use efficiency. Under current circumstances, agricultural water management will maintain the unknown water-efficiency level and farmers will have weaker incentives to make efforts for more efficient practices. A continuous knowledge-exchange is necessary so that all relevant stakeholders can share greater responsibility across the entire water-supply chain. On this basis, more water-efficient management could combine wider environmental benefits with economic advantage for farmers.

Keywords: Innovative irrigation practices; Water-efficiency; Water users organisations; Shared responsibility; Knowledge-exchange system

1. INTRODUCTION

Irrigation systems have been under pressure to produce more with lower supplies of water. Various innovative practices can gain an economic advantage while also reducing environmental burdens such as water abstraction, energy use, pollutants, etc. (Faurès and Svendsen, 2007). Farmers can better use technological systems already installed, adopt extra technologies, enhance their skills in soil and water management, tailor cropping patterns to lower water demand and usage, reduce agrochemical inputs, etc. Water-efficient practices potentially enhance the economic viability and environmental sustainability of irrigated agriculture, without necessarily reducing water usage. To inform such practices, experts have developed various models of water efficiency, yet these are little used by farmers.

Through two case studies in the EU context, this paper will address the following questions:

- When an irrigation area invests in innovative technology, how can its operation help farmers to achieve the full potential benefits together, e.g. an economic advantage, greater water-use efficiency and lower resource burdens?
- Why are innovative technologies often applied in ways which miss the full potential benefits?
- What tensions arise among various objectives and potential benefits?
- How can these difficulties be addressed?

The paper first surveys analytical perspectives on irrigation efficiency - especially the means, incentives and limitations - as a basis to analyse two cases and draw general conclusions.

2. INNOVATIVE IRRIGATION PRACTICES: ANALYTICAL PERSPECTIVES

Innovative irrigation technology is generally promoted as raising water-use efficiency along with multiple benefits, but these remain elusive in practice, as outlined in the first sub-section below. The limitations have fundamental reasons, as outlined in Section 2.2. To address these issues, our case studies are introduced in Section 2.3.

2.1. Practical limitations of water-efficient irrigation technology

EU policy frameworks place great expectations upon technologies to improve water efficiency. The European Commission emphasises 'technological innovation in the field of water, given that water efficiency will be an increasingly important factor for competitiveness' (CEC, 2008). According to the European Parliament, solutions should be found in 'clean technologies that facilitate the efficient use of water' (EP, 2008).

Such technological expectations arise in expert reports on agricultural water use:

Water-efficient irrigation, irrigation on demand and irrigation using brackish water are technologies that will enable the better husbandry of more scarce freshwater resources. Technological developments in respect of irrigation will encompass sensors and communication, intelligent watering systems and high-efficiency delivery mechanisms for water and nutrients, as well as the means of incorporating all of these elements into irrigation 'packages' (EIO, 2011: 25).

Likewise water efficiency can be enhanced by better using current installations and/or by adopting new equipment (WssTP, 2012: 9).

The main European farmers' organisation has likewise advocated technological means to increase water efficiency. In particular this needs 'investments in more efficient irrigation systems, use of new technologies (e.g. soil moisture and canopy sensors) to better match irrigation with plant needs, and good agricultural practices', such as conservation tillage, management of soil fertility and water retention capacity, and scheduling of irrigation during night to reduce evaporation (COPA-COGECA, 2007: 4). The basis for improvement is described as follows:

... water efficiency measures that provide complementary benefits, such as reduced energy needs or other environmental benefits, will also deliver better results. In many Member States, efforts are being made to increase the water storage capacity of soil under agricultural land use. The modernisation of irrigation systems has steadily progressed and water productivity has also improved considerably (COPA-COGECA, 2013: 3)

As indicated above, greater water-use efficiency depends on better agricultural practices alongside extra technology. Yet companies generally promote irrigation technology as if it inherently brings all the benefits (interview, COPA-COGECA, 08.07.13). Improperly managed 'hi-tech' systems can be as wasteful and unproductive as poorly managed traditional systems (Perry et al., 2009). When incorrectly applied, irrigation technology 'can cause losses arising on investments made by farmers, thus decreasing the economic water productivity index and the overall sustainability' (Battilani, 2012).

Beyond a problem-diagnosis of inefficiency, moreover, intensive farming practices can degrade soil and water resources, especially through more input-intensive farming in crops such as maize, vegetables, orchard and vine cultivation:

Intensive arable production is partly responsible for poor soil structure, soil erosion, loss of soil OM [organic matter] and pollution from fertilisers and pesticides.... The expansion of maize cropping and the move to growing winter cereals in particular have contributed to soil erosion even further (Miller, 2007: 44–45).

Such harmful practices have been driven and supported by EU policies. In past decades CAP subsidies have tended to favour crops with high water demands, such as maize, thus increasing the risk of water shortages under climate-uncertain conditions (Garcia-Vila and Fereres, 2012). Either as price-support or area-based, CAP subsidies likewise have ensured the profitability of some water-intensive crops such as cotton which otherwise would be phased out under a market-orientated scenario; likewise water-price subsidies.

In some cases, water-price increases have induced farmers to adopt technology and appropriate practices for conserving water (Caswell and Zilberman, 1985). Yet water-pricing policies often have been ineffective means to reduce water demand (Molle and Berkoff, 2007 and Molle, 2008). Farmers experience rising water prices as an extra penalty. Rather than higher water prices, administrative water allocation or re-allocation lowering the supply often has led farmers to adopt water-efficiency practices (Molden et al., 2010). If agricultural water demand is inelastic, then policies which encourage changes in cropping patterns can be more effective than higher prices (Fraiture and Perry, 2007, Iglesias and Blanco, 2008 and Kampas, 2012).

Inelastic water demand results from farmers' perspectives on water benefits. Water-use efficiency (WUE) and water productivity (WP) are often used interchangeably but have different meanings. WUE specifically means the ratio of biomass produced per unit of irrigation water used, i.e. the sum of transpiration by the crop and evaporation from the soil (Sinclair et al., 1984). By contrast, WP means the ratio of above-ground biomass per unit of water transpired by the crop (Steduto, 2007). Both terms have relevance to farmers' economic goals. WUE interests mainly the water districts or management agencies, while WP interests more farmers and research community. WP better speaks to perspectives linking water usage with production levels and economic benefit (interview, COPA-COGECA, 08.07.13).

Yet even WP remains distant from farmers' perspectives. They generally perceive 'irrigation efficiency' as maximising net revenue rather than saving water (Knox et al., 2012). Policies seek to lower water usage, and river basin managers try to allocate limited supplies, yet water-saving is not a priority for most farmers (Luquet et al., 2005). They manage labour and other inputs to get better economic gains (Molden et al., 2010). Towards that economic aim, most growers make irrigation decisions by relying on subjective judgements, based only on their practical experience and observation (Knox et al., 2012). Consequently, there have been limited benefits from irrigation technology, as well documented in the technical literature; the following examples compare various techniques.

For example, mobile-laboratory evaluations compared the distribution uniformity and irrigation efficiency of various irrigation systems in California. Although microirrigation systems are seen as 'efficient technologies', they were performing less well than traditional surface irrigation methods such as furrows and borders. To gain the extra benefits of such technology, most important is adequate system design, alongside proper installation, operation and maintenance, regardless of the irrigation method used (Hanson et al., 1995).

Howell (2003) and Irmak et al. (2011) reported the attainable application efficiencies for different irrigation methods, assuming irrigations are applied to meet the crops' water needs. Microirrigation has the potential to achieve the highest uniformity (90%) in water applied to each plant, yet poor uniformity and application efficiency can result from various causes, e.g., inadequate maintenance, low inlet pressure or pressure fluctuations, emitter clogging and inadequate system design (Hsiao et al., 2007).

Consequently, microirrigation technology has on-farm efficiencies varying from 0.7 to 0.95 (Howell, 2003).

As another example, a Spanish study compared various irrigation methods via the annual relative irrigation supply index (ARIS), i.e. a ratio of water applied versus water required. It found a greater efficiency of solid-set and drip than surface irrigation. But average annual figures conceal great variations in water applied to a given crop and irrigation efficiency at farm level, partly for lack of adequate knowledge. A remedy would be 'actions to improve farmers' water management via a combination of irrigation advisory services and policy measures' (Salvador et al., 2011: 586).

2.2. Reasons for those limitations and ways to overcome them

Given the above water-efficiency limitations in applying irrigation technology, the literature has outlined some fundamental reasons. They include the following: irrigation equipment is promoted as if the technology per se brings various benefits, farmers seek to maximise net income rather than water productivity per se, innovative technologies can achieve the full potential benefits only through appropriate technical advice, and farmers lack a knowledge-system for anticipating effects of specific irrigation practices or for retrospectively evaluating their irrigation efficiency.

Although research has developed technical scheduling procedures to improve agricultural water management, these have been little adopted, for many reasons.

The one most frequently mentioned by growers is the lack of perceived [financial] benefits relative to their current practices, which they consider adequate. Ease of use and the expenses involved are also important grower considerations (FAO, 2012).

Technical advice on irrigation scheduling is little used at farm level; at most, it helps retrospectively to evaluate seasonal approaches (ibid.).

One obstacle is inadequate knowledge about proper irrigation levels and scheduling over a growing season. Farmers generally lack adequate assistance to develop and adopt better approaches for environmental sustainability, while also maintaining their financial and social objectives (Pereira et al., 2012: 39). For example, sub-surface moisture sensors can improve knowledge about a crop's need for water. But the technology has limitations, so farmers need technical advice to interpret the measurements; for example, 'soil humidity sensors are still neither easy to handle nor reliable' (WssTP, 2012: 33). Moreover, these sensors are not well adapted to all soil types; their installation and maintenance requires the employment of specialised technical staff. The same is true for the canopy sensors, whose proper application is limited to some crops and during specific growing stages, periods of day and climatic conditions.

Improvements in irrigation practices depend on quantitative knowledge of farmers' current practices in relation to actual and potential crop water use:

Any effort to improve water use efficiency needs to start with the assessment of the actual and attainable efficiencies for the given situation, as quantitatively as possible. This information is fundamental for making rational improvements aiming at raising the overall efficiency to the attainable level (Hsiao et al., 2007: 228, 218).

But such information is rarely available to farmers.

Such difficulties arise for water-management improvements through expert systems. Decision Support Systems (DSS) have aimed to improve crop water use efficiency at farm and water basin scale, but few are widely applied, given the necessary specialised skills. For a DSS to be successful, the key elements have been: giving farmers a simple, timely, user friendly, free-of-charge, informative system helpful to decide how much to irrigate in everyday practice; tailoring the tools for a large number of crops; calculating the irrigation profitability; and assessing the economic benefit, especially its relevance to the next irrigation. Such benefits have been demonstrated by the Irrinet project in Italy's Emilia Romagna (Battilani, 2012). Thus more reliable information systems and expert capacity are necessary to guide farmers in using water more efficiently (Battilani, 2013). This exemplifies the broader need for farmer training and education in order to improve modern irrigation management.

As a way forward in the UK, expert support has been recently linking farmers' responsibility, economic benefits and practical knowledge. A 'pathway to efficiency' improves the irrigation network, alongside better practices of soil and water management, e.g. by monitoring whether the right amounts are used at the right place and time. 'Using financial criteria for water efficiency rather than an engineering one appears a sensible approach when assessing irrigation performance at the farm level, since any managerial (e.g. scheduling) and operational (e.g. equipment) inefficiencies associated with irrigation are implicitly included in the assessment' (Knox et al., 2012: 3). In particular, 'On-farm water auditing and benchmarking have the potential to provide useful information to farmer decision making, with respect to identifying operational and management changes to improve irrigation system performance and water productivity, and evaluating potential investments in new technology (and advanced practices) or infrastructures' (ibid: 7).

Such approaches have addressed various obstacles to water-efficiency measures. To exploit the full technological potential requires a broader dissemination of their benefits, specific training of farmers, and coupling properly-designed technological solutions with more precise operational practices to benefit farm economic performance (e.g. Tollefson and Wahab, 1994). In particular, advisory-extension services have enhanced irrigation practices which better fulfil potential benefits of irrigation technology (Hergert et al., 1994,Benham et al., 2000, Ahearn et al., 2003, Genius et al., 2014, Parker et al., 2000 and Gold et al., 2013).

Beyond the farm level, greater resource efficiency also depends on shared responsibility among stakeholders, according to the World Business Council for Sustainable Development:

Business undoubtedly has many opportunities to increase its eco-efficient performance and thereby to help de-couple use of nature from overall economic growth… Establishing framework conditions which foster innovation and transparency and which allow sharing responsibility among stakeholders will amplify eco-efficiency for the entire economy and deliver progress toward sustainability (WBCSD, 2000: 6–7).

Analogous issues arise for service-oriented irrigation schemes, designed so that farmers can flexibly obtain water at their convenience, e.g. through on-demand delivery schedules. Here responsibility has institutional complexities. For example, a water users' organisation (WUO) bears largely fixed costs, as well as somewhat variable energy costs from drainage, excess water application, reuse, disposal, etc. If a WUO or water district relies on gravity-fed water conveyance and delivery systems, then its costs do not vary according to water-volume delivery. In such contexts, if farmers decrease water use, then the WUO must increase water prices to recover its fixed costs. Facing higher water prices, farmers may increase groundwater pumping, thus abstracting more water from aquifers, while distancing their individual practices from any group responsibility. Paradoxically, fostering greater water-use efficiency can generate a financial, environmental and institutional problem.

Given those difficulties for water-efficient techniques, their effective adoption depends on several enabling conditions, especially a policy and institutional context aligning incentives of producers, resource managers and society. Significant synergistic effects can emerge when water-efficiency practices are combined with other agronomic practices such as maintaining soil health and fertility, controlling weeds and avoiding diseases (Molden et al., 2010).

2.3. Methods and sources: EcoWater project

The above issues and earlier questions have been explored through two case studies of service-oriented irrigation schemes within a larger EU-funded research project, EcoWater (see Acknowledgements). It develops a methodology for assessing eco-efficiency at the meso level. The latter is defined as interactions among heterogeneous actors, e.g. between water-service users and providers. As generally understood, eco-efficiency means a ratio between economic advantage and resource burdens, as a basis to evaluate past or potential changes in a system.

The project uses eco-efficiency indicators to evaluate potential innovative practices including technology adoption. The project aims to: assess various options for innovative practices within a specific system; analyse the factors influencing decisions to adopt such practices; and improve understanding of the socio-technical dynamics that influence such decisions.

In the project's two agricultural case studies, farmers and/or their organisations have already invested in water-efficient technology, going beyond state subsidy alone. The irrigation distribution systems were designed for on-demand water delivery. SCADA technology at hydrants allow farmers to abstract water on demand any time and charges them according to a volumetric tiered water pricing. Each case-study area has strong stimuli for farmers to improve water efficiency, yet the full potential benefits of the technology investment were not being realised, for reasons analysed in the next two sections.

3. SINISTRA OFANTO CASE

Dating from the 1980s, the Sinistra Ofanto irrigation scheme is among the largest multi-cropped irrigated areas in Italy. It is located in south-eastern Foggia province within the Apulia region. Irrigation is crucial for the region's agricultural production and income, but it also generates resource burdens. Nearly 18.5% of Apulia's agricultural area is under irrigation; consequently, irrigated crops have contributed 69% of the total value of regional agricultural production, recently quantified as 3.8bn Euros (Fabiani, 2010). The entire study area is characterised by a high number of small land-holdings with intensive, market-oriented practices. The main crops are vineyards, olives, vegetables and fruit orchards (in descending order). The pedo-climatic conditions are favourable for intensive cropping, but profitable farming is strongly dependent on irrigation, due to the scant rainfall and its uneven distribution across the year.

The Sinistra Ofanto system commands an area of 40,500 ha stretching along the left side of the Ofanto River, of which 38,815 ha are irrigable lands and 28,165 ha are serviced with irrigation distribution. Designed and constructed for pressurised on-demand delivery schedule, the irrigation system is managed by a large water users' organisation (WUO), the *Consorzio per la Bonifica della Capitanata* (CBC, 1984 and Altieri, 1995). The system diverts water from the Ofanto River and supplies it to growers both by gravity and lifting/pumping, ensuring a pressure head of at least 2 bar at each hydrant to enable farmers using micro-irrigation methods.

The system is already equipped with modern technologies to deliver and use water efficiently. From the diversion structure on the Ofanto River, water is conveyed to the Capacciotti reservoir through concrete-lined canals and pipe conduits, along which the flow regulation devices are downstream-controlled, thus manually or automatically adjusted through calibrated control devices enabling Supervisory Control and Data Acquisition (SCADA). The Capacciotti reservoir supplies seven concrete-lined storage and compensation reservoirs equipped with downstream-control flow regulation devices that adjust inflows and outflows to feed the district's piped distribution networks based on the downstream water demand.

PVC buried pipes comprise the open-branched distribution networks. Each sector's inlet has a control unit, equipped with flow and pressure metering-control devices. Water is supplied to farms on demand by means of multi-users' electronically-fed hydrants that control and regulate the deliveries, as well as the discharges demanded and thus flowing in the pipe distribution network. These technologies installed along the main infrastructure help keeping conveyance and distribution losses within 5–10% of the total water abstracted from the Ofanto River, as reported by the WUO's engineering staff.

Although the main water supply is surface water, during recurrent water shortages farmers pump groundwater from medium-depth (100–150 m) aquifers, especially since the late 1990s (Portoghese et al., 2013). Furthermore, studies found qualitative degradation of groundwater resources, most likely resulting from seawater intrusion into the coastal aquifer and to deep percolation of pollutants, such as fertilisers and pesticides, from intensive farming activities. Given the urgent need to assess these processes and to avoid their adverse environmental impacts, what are the prospects for water-efficiency improvements of irrigated agriculture in the Sinistra Ofanto area?

3.1. Irrigation patterns and resource burdens

The water users organisation (WUO), Consorzio per la Bonifica della Capitanata (henceforth the CBC), is the main irrigation management agency.

It is responsible for all the sequential steps along the agriculture water supply chain, i.e. abstraction, conveyance, storage, distribution and final water delivery to farm gates. Established in 1933 by a national law of public interest, the CBC is by statute a non-profit organisation; it bears all the costs for performing its functions, and these costs are recovered through the water tariffs paid by farmers.

The CBC enforces the principle of solidarity among the different service areas. Even though the costs for supplying irrigation water differ significantly among areas supplied by gravity and by pumping, the tariff structure does not make such a distinction. Rather, as a tool to manage water use, water fees vary according to demand: volumetric tiered water tariffs progressively increase with the seasonal cumulative volumes withdrawn by each farmer. This structure is enforced through individual water metering at the delivery points; all farm hydrants are equipped with an electro-mechanical delivery device allowing the supply of water only to authorised users and storing information of each irrigation event. Besides simplifying the network operations, this technology proved to be very useful for accurate monitoring and control of water distribution, and for achieving better understanding of the irrigation management practices followed by farmers, especially through the possibility to retrieve and analyse historical data series (Zaccaria et al., 2013).

As an irrigation service provider, the CBC is composed of irrigation service users, i.e. farmers. In performing its daily activities, the CBC attempts to reconcile objectives which may be in conflict. Its technical and administrative choices aim to achieve high water-distribution efficiency in order to maximise the economic benefit to farmers. It aims to improve water distribution and use—at the farm level, through an effective operation of the delivery network, and at field and crop level through the technical support to growers aiming at improved water management skills (ibid). Technical support to farmers was effective in the 1980s–1990s but has declined in the last decade, due to WUO budget constraints and lower revenues from Italy's farm activities.

Irrigation water demand is driven mainly by farmers' perceptions, by the climatic conditions, and by the economic value of crop yields and production factors. Even beyond periods of water shortage, in some areas farmers pump groundwater in order to avoid the following problems: (i) the restricted-flow demand delivery-schedule that prevents the quick completion of irrigation cycles in medium-large farms, (ii) the restricted-frequency demand-delivery often imposed by the CBC during water shortage periods, (iii) the need to arrange water withdrawals with neighbour farmers supplied by the same hydrants, or (iv) the tiered water fees enforced by the CBC. Also, many farmers still perceive groundwater pumping as somewhat cheaper than water supplied by the CBC, even though the contrary was shown by economic analyses (e.g. Portoghese et al., 2013).

As a more fundamental problem, both the farmers' perception and the CBC's analyses ignore the ecological costs of groundwater degradation and remediation. The CBC accepts no responsibility for water-management practices beyond the farm gate. From the growers' standpoint, groundwater pumping aims to increase and/or stabilise the economic benefits of farming activities. Often farmers combine surface water and groundwater for various reasons such as to maximise crop yields and farm net benefit, or to minimise the seasonal water fees payable to the CBC, or to prevent yield reduction arising from high salinity in the groundwater during peak-demand periods. However, this conjunctive use of surface and groundwater is based solely on farmers' economic and technical considerations, regardless of environmental burdens such as aquifer depletion and degradation. Furthermore, fields close to the river banks are often irrigated by growers with water pumped out the river. In all these situations, return flows may result from run-off through the drainage networks, as well as from percolation through the soil profile, finally reaching the downstream reaches of the river, wetlands or the aquifer.

Farm activities generate various pressures on land and water resources, including quantitative depletion and qualitative degradation, especially biodiversity loss in farmland and in the natural environment. This harm has several sources: (i) intensive farming and tillage practices, (ii) fertilisers and pesticides application on cultivated fields, (iii) water abstraction from the Ofanto River, (iv) return flows of degraded water to downstream wetlands and aquifers, (v) over-drafting of groundwater, (vi) salinity build-up in cultivated soils, (vii) energy consumption for water pumping, and (viii) increased CO_2 emissions from the energy usage related to pumping, transport, machinery, etc.

Relative to those ecological problems, much greater impetus for innovative practices comes from recurrent scarcity of water supply and the prospect of even greater future scarcity and uncertainty. Those problems in turn result from high water-demanding crops and from irrigation scheduling practices. Such decisions are often based solely on farmers' perceptions; their systems and practices are not monitored to assess the actual performance and efficiency achievements. No systematic technical support is available to growers for their daily or seasonal irrigation planning and scheduling.

Moreover there is detailed evidence of water-use inefficiency at farm level. According to a study of farmers' irrigation practices in a nearby irrigated area with similar features, there were often mis-matches between crops' water demand and irrigation applications on several occasions during the season. Although the overall seasonal applied irrigation depths may match

Table 1. Technologies, impacts and necessary improvements.

Technology	Level of application	State of implementation	Action	Impacts	Needs and Improvments
Multi-user electronic hydrants	Distribution network/hydrants	Installed and operated	Supply water to authorised users Control, record and store irrigation events Bill farmers according to used volumes Implement tiered water tariff Programmable to supply water during specific time slots	Simplify water distribution and water billing Reduce WUAoperational cost Responsible water use by farmers Reduce water demand	Data transmission Low-pressure alert High-salinity alert Integration with IAS
Variable-speed pumps	Pumping plant	Recently installed S operated	Adjust their operation dynamically Supply the discharges and pressure heads required by the network	Energy saving Reduce WUO operational costs	Data recording and transmission Accurate system analysis Tuning operational regulation Improve energy-efficiency
Shift to improved irrigation methods (sprinkle to mini-sprinkle and trickle)	Farm irrigation systems/cropped plots	Recently implemented by some farmers	Reduce soil wetting Reduce soil evaporation losses Reduce energy requirements	Reduce water demand Energy saving Reduce irrigation cost to farmers Reduce overall water abstraction	Better visualisation of benefits Technical support and extension Irrigation advisory service Improve application efficiency

Sub-surface drip irrigation (SDI)	Farm irrigation systems/cropped plots	Very limited	Reduce/minimise soil wetting Reduce application losses Increased WP	Reduce water demand Energy saving Reduce irrigation cost to farmers Reduce overall water abstraction	Precision operation Better visualisation of benefits Improve irrigation scheduling Improve application efficiency Technical support and extension Irrigation advisory service
Regulated deficit irrigation (RDI)	Cropped plots	Very limited	Increased WUE Increased WP	Reduce water demand Energy saving Reduce irrigation cost to farmers Reduce overall water abstraction	Soil and plant water status monitoring Precision irrigation Better visualisation of benefits Technical support and extension Irrigation advisory service

a crop's water demand, farmers often under-irrigate during the early crop stages and over-irrigate during later stages; many choose inadequate timings and application depths. Such inadequate applications may be combined with uneven in-field water distribution, often due to the average low uniformity of irrigation systems—especially when not properly designed, evaluated and maintained; consequently, the farm may have up to 20% lower crop yields and income, along with inefficiencies between 20 and 40% due to excessive water applications (Zaccaria et al., 2010). As the main reason for the mismatch, irrigation scheduling practices are based only on farmers' perceptions and experiences; missing is status monitoring of soil or plant water, use of ET-based irrigation scheduling, or any other quantitative techniques (ibid). This study confirms a general problem of water-inefficient practices, as also found in the wider technical literature (e.g. Hanson et al., 1995, Hanson et al., 1996, Burt, 2004 and Salvador et al., 2011; see Section 1).

3.2. Innovative practices for stakeholders' consideration

Resource-efficiency could be enhanced by properly utilising several innovative technologies and practices. As listed in Table 1, several feasible options are already installed and implemented in the Sinistra Ofanto area, i.e. along the water conveyance and distribution system or on some progressive farms, but require either some refinements or significant operational improvements to gain their full economic and environmental benefits.

Investment costs for the installed technologies were paid by the WUO either from budget surpluses (in years when water demand and delivery were high), or by funds obtained from different sources, and by some individual farmers at the farm level. Their maintenance and operational costs are paid by growers within the regular water fees. The improved irrigation methods, systems and practices are implemented only to a small extent by some farmers.

The first two technologies were recently installed along the distribution networks to ease operation and flow control, and thus increase its efficiency, but they still need further testing and tune-ups to fully exploit their potential. For example, multi-user hydrants could be hooked up with soil moisture sensors or plant–water status sensors to schedule irrigation applications on the basis of needed or user-defined thresholds and frequency. Also, they could be set to cease water withdrawals and irrigation when the pressure-head values at delivery are insufficient to achieve the target level of efficiency and distribution uniformity. Moreover, relevant datasets from the field (on irrigation timings, flow rates, pressure heads, volume and duration) could be automatically transmitted to a central WUO workstation, thus easing the data-gathering necessary for billing farmers, for evaluating the irrigation delivery service provided and growers' irrigation scheduling practices. These further adjustments of the available technologies could also enable the performance evaluation and benchmarking of irrigation at on-farm and project level.

As indicated by those examples, upgraded technologies alone do not ensure the anticipated resource- efficiency improvements. Compared to less technology-intensive systems, the past investments incur extra costs but bring uncertain gains. The full potential benefits depend on adequate management practices, in turn dependent on better expertise, skill-gap assessment, auditing monitoring and evaluation, control systems and extension programmes, including mission-oriented research and outreach.

The above technology options (in Table 1) were presented at the EcoWater project workshop, held in the case study area and attended by diverse stakeholders involved in the Sinistra Ofanto water-system value chain. These included: the River Basin Authority, the WUO, the Apulia Regional Administration, the regional Department of Water Resources, the Ente Irrigazione, ARPA Puglia, CNR-IRSA (Water Research Institute), University of Bari, etc. Stakeholders, decision-makers and experts were asked to share their visions about water management in the area. The discussion identified several issues, emphasising the need for measures to protect groundwater.

As many comments there highlighted, in the last three decades contradictory interests have generated conflicts over water allocation and use. Surface-water availability for agricultural use has been reduced by several factors: expansion of the command areas, incentivised partly by irrigation systems and by regional policies; greater water demands of the municipal, industrial and tourist sectors; and adverse environmental-climatic changes. Participants mentioned several solutions, in particular: monitoring and control of water use should be strengthened throughout the Apulia region, especially in the study area, in order to avoid aquifer over-exploitation and uncontrolled withdrawals from water courses; deficit irrigation management, already practiced in the study area by some innovative growers, offers a feasible option to enhance water-use efficiency and productivity; and, finally, farmers should share their personal experiences and knowledge in order to help others to use irrigation water in a sustainable and efficient way (reported in Ecowater, 2013).

In addition, CBC (local WUO) representatives emphasised the importance of using non-conventional water sources as an additional supply, especially the re-use of treated wastewater (TWW) as a valid alternative to aquifer exploitation for irrigation purposes. TWW re-use has several barriers: stringent quality criteria set by regulations for water reuse in agriculture; the high cost of treatment and water conveyance from the treatment plants to agricultural areas; and inadequate research dissemination, even disinformation; and reluctance by farmers. Consequently, TWW re-use should be fostered by several means – lowering the water-quality restrictions, installing the latest technologies available on the market for wastewater treatment to reduce costs of treatment plants, and broadly disseminating research findings to growers, especially through extension activities and farmers' advisers – according to the WUO representatives (reported in Ecowater, 2013). No stakeholder raised concerns about aquifer characterisation, safe-yield assessment, groundwater recharge or agrochemicals' fate in soil and aquifers, or farm-level performance assessment of agricultural water management practices.

As emerged during the stakeholder discussion, some future visions may be based on doubtful assumptions. In particular the CBC staff and farmers' representatives assumed that Sinistra Ofanto farmers already achieve high irrigation and water-use efficiency, simply on grounds that they use microirrigation methods. From this assumption, there would be little scope or incentive to growers for further improvement in farm-level water management.

From CBC technical reports over at least a decade, no recent information is available on whether the farm irrigation systems are properly designed, installed, operated and maintained—nor on whether irrigations are adequately scheduled and conducted by growers. Likewise the WUO has not recently fostered irrigation systems evaluations regarding the actual application efficiency and distribution uniformity achieved by growers, or WUE/WP for the various crops and irrigation methods, on the basis of quantitative measurements. Agricultural extension activities have been significantly limited in the last decade by budget constraints. So the CBC lacks an empirical basis for its efficiency assumptions about current performance of on-farm water management. In all those ways, the CBC's practical responsibility ends at the farm gate.

Moreover, responsibility for water-supply problems is displaced onto public-sector agencies, especially for additional water supplies via wastewater reuse. This option is simpler and perhaps more profitable for farmers and the CBC, especially if public bodies carry the main financial burden. Water reuse involves complex issues needing careful investigation and management, in particular: different water qualities may be required, contingent on the final water uses. Water that needs to be buffered or stored for seasonal reuse (e.g. groundwater banking or surface storage) may have other quality requirements than water used for a direct agricultural reuse. And there are potential risks to health and environment (WssTP, 2013: 19–20).

4. MONTE NOVO CASE

In Portugal's Alentejo the Monte Novo irrigation perimeter provides abundant water for a rising number of farmers, served by the larger Alqueva reservoir and irrigation project. Dating from 2009, the scheme set initially low water prices to incentivise irrigation by farmers—but without their involvement in discussing implications for crop choices, irrigation practices, farming profitability and expected income. The highly-discounted water price has incentivised crops with high water demands as well as agrochemical inputs, thus creating pollution problems. As mandated by law, the water price will rise significantly towards full-cost recovery by 2017, thus potentially stimulating practices which need less water. What are the prospects for water efficiency improvements?

4.1. Irrigation patterns and future price rise

Since the 1960s Portugal has had plans to irrigate relatively arid areas of the Alentejo, especially to increase employment and livelihoods. Spain's 1993 National Hydrologic Plan was proposing to abstract more of the shared water from the adjoining region, partly on grounds that Portugal was not using the water, but Portugal objected. To avoid such trans-border conflicts, the 1998 Albufeira Convention aimed to protect freshwater and groundwater, as well as the sustainable use of shared water resources, in the context of the negotiation process for the Water Framework Directive (Maia, 2000 and Pulwarty and Maia, 2013). Under the Convention, for example, Spain guarantees annual stream flows to Portugal in normal years (Costa, 2003).

The Albufeira Convention facilitated Portugal's Alqueva Multipurpose Project—the Empreendimento de Fins Múltiplos de Alqueva (EFMA). The EU Structural Funds were expected to pay more than half the investment cost (Vergés, 2001). The 1993 EFMA plan combined a reservoir, hydroelectric dam and irrigation networks, thereby helping to justify Portugal's claim on the water. The project aimed to promote the development of a poor, deprived region through irrigated agriculture, electricity production and tourism. Together these were meant to ensure the sustainability of the project in its early phase (Santos et al., 2011).

From the start the EFMA was criticised as a hydrodinosaur—for accepting engineers' assumptions about future water needs, requiring enormous electricity to transport water and so raising the full cost (Costa, 2003). Before EFMA's operation, maize production had greatly risen alongside greater dependence on CAP subsidies for economic viability. The Agriculture Minister hoped that the Alqueva would irrigate industrial crops such as tomato and sugar beet (Vergés, 2001: 7). By using cheap water from the Alqueva dam, some Alentejo farmers were expecting to capture greater payments from the CAP 1st pillar. So they opposed mid-term reforms decoupling such payments from production levels; so did the Portuguese government (Costa, 2003: 26).

According to its promoters, the Alqueva system will foster significant regional development in Alentejo, both in social and economic terms, due to the strategic water reserve and the associated hydraulic infrastructures. The main objective is to guarantee an adequate volume and quality of water to users, especially irrigation water. Further objectives include the energy hydropower production, urban water supply, ecological conservation and recreational uses (EDIA, 2011).

With its geographical topography and high elevation, the Alqueva scheme was an expensive investment (Costa, 2003), especially relative to the small number of farmers who initially joined the scheme. This arrangement resulted in a high unit cost for the water supply. This has high political-economic stakes, given that full-cost recovery of irrigation systems is required by the Water Framework Directive (EC, 2000).

Irrigation delivery service is provided by the public company EDIA (Empresa para o Desenvolvimento das Infraestruturas de Alqueva), the agency responsible for the Alqueva project development and exploitation. Water is abstracted from the Alqueva reservoir and transported through a network of canals and ducts, from primary network to secondary network through hydrants to farmers. Within the larger Alqueva project, the Monte Novo irrigation perimeter is located in Alentejo district, near Évora municipality. The perimeter provides water for irrigation to an area of at least 7800 ha, while the Alqueva Project will have a total 115,000 ha expected capacity by 2015.

In the Guadiana region including Monte Novo, the tertiary sector has 76% of gross value added (GVA) and 67% of employment; the most important sources are tourist accommodation, catering and public services. The agricultural sector has 10% of the region's GVA and about 15% of the jobs; average farm size is large (55 ha). These characteristics, associated with the high proportion of farmers in a company structure, indicate the region's high potential to develop agricultural activities, especially with the full operation of the Alqueva project. But its agricultural activities have a very low competitiveness and a low productivity per unit area; incomes are greatly supported by public subsidies, which comprise about 65% of the total gross margin (ARH-Alentejo, 2011).

In the study area, the major irrigated crops are olives, maize, arable crops (mostly pasture) and horticultures (mostly tomatoes). Farm size ranges from less than 50 ha to more than 500 ha. The largest areas belong to important multi-crop farms, such as the ones owned by the Fundação Eugénio de Almeida (FEA), or the olive farms owned by Olivais do Sul (ODS); together they comprise more than 30% of the irrigated area in Monte Novo irrigation perimeter.

Fertilisers and pesticides are applied to all crops, especially tomatoes and maize. Greater irrigation stimulates run-off, leach-outs of fertilisers and pesticides and soil erosion. Together these practices reduce soil organic matter, soil fertility and its capacity to retain water, in turn increasing potential irrigation inefficiencies and need for better water management for the same cultivation level as before. Indeed, land degradation and nutrient depletion significantly constrain opportunities to increase water efficiency (Molden et al., 2010 and IST, 2013).

In 2009 the WUO Associação de Beneficiários de Monte Novo (henceforth ABMN) was established, representing all farmers connected to the Alqueva water distribution system from EDIA. According to the regulatory Decree Law 84/82, the Association has formal recognition by the Ministry of Agriculture, Trade and Fisheries. The ABMN has the role to promote the administration of constructing the hydro-agricultural development. Nonetheless EDIA has carried out the management and operation of the irrigation perimeter, as well as the investments in constructing the irrigation network.

Water prices are differentiated according to the pressure heads provided at farm-gate delivery. Prices are lower in low-pressure blocks, where water is conveyed and distributed by gravity, generally to larger farmers; irrigation inside farms is within farmers' management. Prices are higher in high-pressure blocks, which are supplied by pumping stations; these reduce the need for farmers to bear the costs of farm-level booster pumps. However, in the high-pressure blocks some large farms need their own investment in irrigation equipment and distribution infrastructures. In some cases farmers already invested in important irrigation infrastructures prior to the construction of the Monte Novo irrigation perimeter. In both cases a higher water price is imposed on important farmers who will not necessarily benefit from higher pressure heads at hydrants (Rita and Capelo, 2005). Moreover, the difference in water prices between low-pressure and high-pressure blocks is insufficient for covering the necessary investments for constructing the main irrigation infrastructures within low-pressure farms' areas, as well as the corresponding exploitation costs, mainly energy and maintenance (Froes and Rodrigues, 2005).

Given those high costs, full-cost recovery would impede a full transition from rain-fed to irrigated agriculture. To attract farmers into the scheme, a 2010 law set the initial water price at only 30% of full cost. This initial price shifts more of the total cost onto the supplier, while incentivising maize cultivation, which demands relatively large amount of water but gains a higher market price. According to the ABMN (Monte Novo's WUO), the discount price gives farmers 'a certain hope', but many have difficulties to obtain credit, which is necessary for farm-level investment to electrify the cultivated zones (Silva, 2010).

As originally set by law, the low water price will be increased by 10% per year towards full-cost recovery by 2017 (Nuncio and Arranja, 2011). This legislative requirement has become contentious. At its full-cost price, water would comprise 13% of the total cost of maize production. Through their WUOs, farmers have been lobbying the authorities to delay the price rise. According to the Federação das Associações de Agricultores do Baixo Alentejo (regional federation of farmers' associations), the higher price will not allow farmers to be economically competitive, especially for a large proportion of maize cultivation, estimated as more than 36,000 tonnes aimed at export markets. According to the former Agriculture Minister, however, the higher price will be competitive with nearby regions, as an adequate basis for more farmers to join the scheme (Maneta, 2010 and Rosado, 2012). The future increase could stimulate water-conserving measures and alternative cropping patterns which would be more profitable and/or need less water, while also reducing other resource burdens. But continuous support measures would be necessary for growers to evaluate and implement such options.

4.2. Innovative practices for stakeholders' consideration

The on-demand water-delivery system was meant to incentivise water-efficient practices by farmers and thus to reconcile tensions between various objectives of the Alqueva project. Such a strategy put great expectations upon a technological system. At the irrigation-perimeter level the eventual implementation of innovative practices is meant to be supported by the ABMN (WUO), with investment costs if necessary to be covered from budget surpluses. On the other hand, the farm-level water-use improvements are dependent on farmers' own investment or eventually on government subsidies fostering efficiency increases. Some technologies have been applied at farm level, e.g. sub-surface irrigation at FEA and ODS farms, but only to a small extent. Buried drip-irrigation systems need even more careful design, monitoring and operation than surface-drip or other micro-irrigation systems. Beyond the technological investment, greater application would depend on applied research, demonstration activities, outreach, farmers' training and skills development.

The study team assessed various options for lowering input demands (of energy, water, agrochemicals) or for increasing farmers' income per unit water used (Ecowater, 2012), in two main categories as follows. At the irrigation perimeter level, i.e. the distribution network:

- tiered volumetric water tariffs according to actual water use by growers;
- tiered water tariffs according to timing of withdrawals and energy costs, e.g. lower rates at night-time; and
- pressure head delivery.

And at the farm level:

- drip irrigation, reducing water evaporation (especially relevant to maize);
- sub-surface drip irrigation, minimising soil evaporation and facilitating mechanical weed-control or conservation tillage or minimum-tillage methods (especially relevant to vineyards);
- super-high density olive orchards;
- variable-irrigation practices, e.g. through regulated deficit irrigation; and
- alternative crops demanding less water.

The study team presented the above options at an April 2012 workshop. It was attended by representatives and experts from relevant local bodies. They included: EDIA, farmers' representatives (FEA and ODS), the ABMN (WUO), ARH-Alentejo, and the Centro Operativo de Tecnologia do Regadio (COTR), an advisory body for scientific research regarding agricultural development.

Stakeholders' comments converged around the following points:

- All the proposed technologies could add value to the Alqueva scheme.
- Farmers are interested in any technological configuration that might increase their profit margins, which are currently low.
- Given the high investment costs of the irrigation scheme, a successful operation is important in order to lower the unit cost of water through access to more growers (Ecowater, 2013).

Discussion focused on the knowledge lacking for farmers to minimise irrigation intensity and to conserve soil resources. In particular participants made these comments:

- Irrigation intensity must remain within the carrying capacity of the soil (infiltration rate and water-holding capacity), especially in order to prevent surface run-off, leach-outs and erosion—significant environmental impacts that must be taken into consideration.

- Without adequate knowledge for such judgements, farmers may intensify resource usage, thus increasing costs, leach-outs and soil erosion.
- Root-zone soil moisture conditions are not measured to identify in-field variability and vulnerable areas, so most farmers base decisions on their past experience and daily observations of farm conditions.
- Although a network of meteorological stations already exist in the area, relevant information is not available for irrigation planning and scheduling purposes; farmers' access is still under development (Ecowater, 2013).

Regarding the above point about the soil's water-capacity, a COTR expert emphasised soil organic carbon (SOC) as an environmental indicator and objective. Greater irrigation flows undermine the soil's capacity to retain water and nutrients, while also increasing GHG emissions. SOC conservation is more difficult for annual arable crops than for permanent pasture. To avoid further SOC loss and to enhance its role, farmers could incorporate crop residues into the soil and/or choose crops which need less irrigation and chemical inputs. Before any technological change, decision-makers should evaluate the implications for SOC, he argued. This issue was not taken up by other stakeholders at the workshop.

The workshop discussion also considered whether the cultivation of organic crops could be an alternative option, along with bio-labelling to gain a higher market price. According to the ODS representative, farmers would use organic cultivation methods if they could be convinced that their profit would increase. But owners of small farms would not be easily convinced (Ecowater, 2013). Indeed, small-scale growers lack an advisory service and systematic support for linking organic methods with higher-value markets. This institutional gap illustrates wider difficulties for farmers adjusting to new challenges and gaining the full potential benefits of the EFMA.

5. DISCUSSION AND CONCLUSIONS

Innovative irrigation practices can enhance water efficiency, gaining an economic advantage for farmers while also reducing environmental burdens. Water-efficient methods and better irrigation scheduling could also integrate water and nutrient management, thus minimising agrochemical runoff and leaching problems. To help fulfil this potential, experts have developed various models of water efficiency and environmental benefits. Yet these models are little used for irrigation scheduling; at most, they help retrospectively to evaluate seasonal approaches (FAO, 2012).

In that generally adverse context, our two case studies of irrigation schemes investigated options, stimuli and difficulties to improve water-efficient practices. In both cases, individual farmers and/or their organisations already made significant investments in irrigation technologies, but their implementation has not been systematically evaluated for effectiveness through better knowledge—regarding crops' water use, soil-moisture conditions, irrigation scheduling techniques, crops' yield response to different irrigation management strategies, etc. Farmers there have no formal responsibility to demonstrate efficient water use (unlike in the UK; Knox et al., 2012). Consequently, farmers pay higher water prices yet do not obtain the full potential benefits through water-efficient practices. Reasons why have been investigated by exploring perspectives of water users organisations (WUOs) and relevant agencies, especially through multi-stakeholder workshops.

The two cases have strong stimuli for more water-efficient practices. In the Sinistra Ofanto case, a main stimulus is the prospect of greater water scarcity and erratic supply, leading some farmers to rely on groundwater, which may progressively become more scarce, over-drafted, more expensive due to higher pumping costs, degraded and/or saline. In the Monte Novo case, a water-abundant system, a main stimulus is the statutory rise in water price towards full-cost recovery; this would weaken the commercial viability of some crops with high water demand, such as current maize cultivation.

Anticipating such future difficulties, stakeholders have discussed options for better practices. In each area, stakeholder representatives attended our project workshop on how farmers' economic advantage could be combined with environmental benefits. Despite the stimuli for more water-efficient practices, however, actors' responsibility has been somewhat displaced onto hypothetical prospects. In particular, for the two areas:

At the Sinistra Ofanto workshop, stakeholder representatives emphasised problems of water scarcity and aquifer salinity. Their comments suggested various possible measures such as region-wide control and regulation, farm-level practices and farmers' knowledge-exchange systems. At the same time, however, the WUO displaced responsibility onto the prospect of additional supply from wastewater reuse; this would depend on significant institutional, policy and technical changes, with an uncertain timescale and cost.

At the Monte Novo workshop, stakeholder representatives expressed interest in farm-level technological improvements which could improve water efficiency. Yet WUO representatives also emphasised the need to maintain a low unit cost of water. Moreover, the WUO publicly advocates a longer-term low water pricing, which would depend on a legislative change contradicting the country's EU obligations and the original mandate for the irrigation scheme.

In both cases responsibility is displaced, thus complementing WUOs' and farmers' assumption that their agricultural water management practices already have adequate efficiency. This attitude arises by default—from seeing no systematic means or incentive for more water-efficient practices, beyond the modern technologies already adopted. Experts generally have sought to overcome such obstacle through better models of evapotranspiration and efficient water use (e.g. Pereira et al., 2012). Yet such models cannot reconcile stakeholders' different understandings of irrigation efficiency: farmers seek to maximise their economic benefits, while water experts or agencies seek to conserve water (Knox et al., 2012).

Therefore a solution would need a continuous knowledge-exchange with three components:

i. expert scientific knowledge of crops' water needs, their yield-response to water and the actual on-farm versus attainable efficiency (Hsiao et al., 2007);

ii. links between farmers' perspectives, innovative practices and their income benefits (Knox et al., 2012); and

iii. means to lower resource burdens from inputs and pollutants (Miller, 2007).

iv. Relative to such a solution, our two case-study areas have a nominal extension service which lacks the necessary resources, expertise and operational capability in agricultural water management; these services have no external auditing. A knowledge-exchange system does not exist in those areas, so farmers may miss the full potential benefits of the technological investment already adopted. Under current circumstances, irrigation practices will maintain the unknown efficiency level; along with their WUOs, farmers will have weak capacities to make extra efforts for more water-efficient practices, as well as weak incentives to pay for any further investments.

An adequate knowledge-exchange system would depend on greater institutional responsibility for water-related policies and strategies, for example, through an extension service or a functional equivalent through a WUO. This in turn would provide an enabling condition and incentive for all relevant stakeholders (especially WUOs) to share greater responsibility for agricultural water management across the entire water-supply chain, including farm-level practices as well as drainage and leach-out management. On this basis, more water-efficient practices could combine wider environmental benefits with economic advantage for farmers.

ACKNOWLEDGEMENTS

This work was supported by the European Commission through the project, 'EcoWater: Meso-level eco-efficiency indicators to assess technologies & their uptake in water use sectors', a collaborative research project of the 7th Framework Programme, grant agreement no. 282882, coordinated by the National Technical University of Athens (NTUA) during 2011-2014, http://environ.chemeng.ntua.gr/ecowater/. We would like to thank two anonymous referees of this journal for helpful comments.

REFERENCES

1. ReferencesAhearn, M., Yee, J., Bottom, J., 2003. Regional trends in extension system resources.In: Agricultural Information Bulletin No. 781. U.S. Department of Agriculture,Economic Research Service, Washington, DC.

2. Altieri, S., 1995. The Sinistra Ofanto irrigation scheme: management and mainte-nance problems. In: Pereira, L.S. (Ed.), Bonifica, vols. 1–2. Bastogi Editrice Italiana,Foggia, pp. 40–47.

3. ARH-Alentejo, 2011. Planos de Gestão das Bacias Hidrográficas–RegiãoHidrográfica 7–Relatório técnico para efeitos de Participac, ão Pública,Nemus–Ecossistema–Agro.ges (River basin management plan—Guadiana Tech-nical report for public participation). Administrac, ão da Região Hidrográfica(ARH) do Alentejo.

4. Battilani, A., 2012. Sustainable Knowledge-based Irrigation Management: TheIRRINET package, _http://ec.europa.eu/environment/archives/greenweek2012/adriano-battilani.html_,

http://ec.europa.eu/environment/archives/greenweek2012/sites/default/files/3-9 battilani.pdf.

5. Battilani, A., 2013. Approaches to Improve the Productivity of Water Use in Irrigation[with diagrams], _http://www.wsstp.eu/files/WSSTPX0001/II 2 %20Adriano%20Battilani.pdf_.
6. Benham, B.L., Linder, G.M., Hinrichs, M.D., 2000. Improving irrigation managementin Nebraska's central Platte valley. J. Nat. Res. Life Sci. Educ. 29, 111–115.
7. Burt, C., 2004. Rapid field evaluation of drip and microspray distribution uniformity.Irrig. Drain. Syst. 18, 275–297.
8. Caswell, M., Zilberman, D., 1985. The choices of irrigation technologies in California.Am. J. Agric. Econ. 67 (2), 224–234.
9. CBC, 1984. Consorzio di Bonifica per la Capitanata, Cinquant'anni di bonifica nelTavoliere. Bastogi Editrice, Foggia.
10. CEC, December, 2008. Follow up Communication on Water Scarcity and Droughtsin the European Union COM (2007) 414 final, [SEC(2008) 3069]. Commission ofthe European Communities, Brussels.
11. COPA-COGECA, 2007. COPA-COGECA Position on Water Scarcity and Drought.COPA (Committee of Professional Agricultural Organisations), General Con-federation of Agricultural Cooperatives (COGECA), Brussels, _http://www.copa-cogeca.be/Menu.aspx_.
12. COPA-COGECA, 2013. Water and Agriculture under a Changing Climate. COPA-COGECA.
13. Costa, L., 2003. A Portuguese Hydrodinosaur: The Hidden Costs of the Alqueva Dam,_http://www.hydrologie.org/hydrodinosaures/download/portugal ang.pdf_.
14. EC, 2000. Directive 2000/60/EC of the European Parliament and of the Council of 23October 2000 Establishing a Framework for Community Action in the Field ofWater Policy. EC.
15. Ecowater, 2012. Deliverable 1.2: Value Chain Mapping of the AgriculturalWater Systems. Ecowater, EcoWater project, _http://environ.chemeng.ntua.gr/ecoWater/Default.aspx?t=238_.

16. Ecowater, 2013. Deliverable 6.1: Synthesis report from the 1st Round ofCase Study Events. Ecowater, _http://environ.chemeng.ntua.gr/ecoWater/Default.aspx?t=238_.

17. EDIA, 2011. Portal de mapas (Map Portal). EDIA, _http:\\www.edia.pt_ (last visit inDecember 2011).

18. EIO, 2011. Water Innovation: How Eco-innovation can Contribute to the Sustaina-bility of Europe's Water Resources. Eco Innovation Observatory.EP, 2008.

19. Addressing the challenge of water scarcity and droughts in the EuropeanUnion. European Parliament resolution of 9 October 2008 on addressing thechallenge of water scarcity and droughts in the European Union (2008/2074(INI).Fabiani, S. (ed.), 2010. Aspetti economici dell'agricoltura irrigua in Puglia.Istituto Nazionale di Economia Agraria (INEA), http://dspace.inea.it/bitstream/inea/520/1/SE5-1164.pdfFAO, 2012. FAO Irrigation and Drainage Paper 66: Crop Yield Response to Water,_http://www.fao.org/nr/water/infores.html_.

20. Faurès, J., Svendsen, M., Turral, H., 2007. Reinventing irrigation. In: Molden, D. (Ed.),Water for Food, Water for Life: A Comprehensive Assessment of Water Manage-ment in Agriculture. Earthscan and International Water Management Institute,London, Colombo (Chapter 9).

21. Fraiture, C., Perry, C., 2007. Why is agricultural water demand unresponsive at lowprice ranges? In: Molle, F., Berkof, J. (Eds.), Irrigation Water Pricing: The Gapbetween Theory and Practice. CABI, Wallingford, pp. 94–107.

22. Froes, J., Rodrigues, P.N., 2005. Distribuic, ão de água em baixa pressão no Alqueva.In: Paper in I Congresso da Rega e Drenagem. Beja, Portugal, Available in_www.cotr.pt/informacao/web2/Papers/48.pdf_.

23. Garcia-Vila, M., Fereres, E., 2012. Combining the simulation crop model AquaCropwith an economic model for the optimization of irrigation management at farmlevel. Eur. J. Agron. 36, 21–31.

24. Genius, M., Koundouri, P., Nauges, C., Tzouvelekas, V., 2014. Information transmis-sion in irrigation technology adoption and diffusion: social learning, extensionservices, and spatial effects. Am. J. Agric. Econ. 96 (1), 328–334.

25. Gold, A.J., Parker, D., Waskom, R.M., Dobrowolski, J., O'Neill, M., Groffman, P.M.,Addy, K., 2013. Advancing water resource management in agricultural, rural,and urbanizing watersheds: why

land-grant universities matter. J. Soil WaterConserv. 68 (4), 337–348.

26. Hanson, B., Bowers, W., Davidoff, B., 1995. Field performance of microirrigation sys-tems. In: Microirrigation for a Changing World, Proceedings of Fifth InternationalMicroirrigation Congress, April 2–6, 1995. American Society of Agricultural Engi-neers, Orlando, FL, pp. 769–774.

27. Hergert, G.W., Penas, E.J., Ferguson, R.B., Cahoon, J.E., Klocke, N.L., Watts, D.G.,Shapiro, C.A., Peterson, T.A., 1994. Extension programming for improving nitro-gen and irrigation management in Nebraska. In: National Symposium onProtecting Rural America's Water Resources: Partnerships for Pollution Solu-tions, Washington, DC.

28. Hanson, B., Bowers, W., Davidoff, B., Kasapligil, D., Carvajal, A., Bendixen, W., 1996.Field performance of microirrigation systems. In: Proceedings of the Fifth Inter-national Microirrigation Congress, Orlando, FL, pp. 769–774.

29. Howell, T.A., 2003. Irrigation efficiency. In: Encyclopedia of Water Science.,http://dx.doi.org/10.1081/E-EWS 120010252, http://www.researchgate.net/publication/43256707

30. Irrigation EfficiencyHsiao, T.C., Steduto, P., Fereres, E., 2007. A systematic and quantitative approach toimprove water use efficiency in agriculture. Irrig. Sci. 25, 209–231.

31. Iglesias, E., Blanco, M., 2008. New directions in water resources manage-ment: the role of water pricing policies. Water Resour. Res. 44 (6), 1–11,http://dx.doi.org/10.1029/2006WR005708.

32. Irmak, S., Odhiambo, L.O., Kranz, W.L., Eisenhauer, D.E. 2011. Irrigation Efficiencyand Uniformity, and Crop Water Use Efficiency, Lincoln, NB: University ofNebraska–Lincoln Extension, http://extension.unl.edu/publicationsIST, 2013. Avaliac¸ ão Ambiental Estratégica dos novos investimentos EFMA relativosà Rede Secundária de Rega, no âmbito do INALENTEJO (Strategic Environ-mental Assessment of new EFMA's Investments). In: Relatório preliminar:Relatório Ambiental. Instituto Superior Técnico (IST), Lisboa, Available onlineat _http://www.inalentejo.qren.pt/images/stories/file980.pdf_.

33. Kampas, A., Petsakos, A., Rozakis, S., 2012. Price-induced irrigation water saving:unravelling conflicts and synergies between European agricultural and waterpolicies for a Greek water district. Agric. Syst. 113 (1), 28–38.

34. Knox, J.W., Kay, M.G., Weatherhead, E.K., 2012. Water regulation, crop produc-tion and agricultural water management—understanding farmer perspectiveson irrigation efficiency. Agric. Water Manage. 108 (1), 3–8.

35. Luquet, D., Vidal, A., Smith, M., Dauzatd, J., 2005. More crop per drop: how to makeit acceptable for farmers? Agric. Water Manage. 76 (2), 108–119.

36. Maia, R., 2000. Portuguese–Spanish river basins: bilateral agreements' evolution andcontext. Water Sci. Technol. 42, 227–233.

37. Maneta, L., abril, 2010. Prec¸ o da água de Alqueva não convence agricul-tores. In: Diario de Noticias, _http://www.dn.pt/inicio/economia/interior.aspx?content id=1540958_.

38. Miller, A., 2007. Environmentally Sustainable Agri-Food Production. EnvironmentalSustainability Knowledge Transfer Network (ESKTN).

39. Molle, F., Berkoff, J., 2007. Water pricing in irrigation: the lifetime of an idea. In:Molle, F., Berkoff, J. (Eds.), Irrigation Water Pricing: The Gap between Theoryand Practice. CABI, Wallingford, pp. 1–20.

40. Molle, F., 2008. Can water pricing policies regulate irrigation use? In: 13th WorldWater Congress, September, Montpellier, France.

41. Molden, D., Oweis, T., Steduto, P., Bindraban, P., Hanjra, M.A., Kijne, J., 2010. Improv-ing agricultural water productivity: between optimism and caution. Agric.Water Manage. 97, 528–535.

42. Nuncio, J., Arranja, C., 2011. Gestão dos aproveitamentos hidroagríco-las de Alqueva. Jornadas Tecnicas APRH, _http://sir.dgadr.pt/conteudos/jornadas aph/apresentacoes/s7/11.pdf_.

43. Parker, D., Cohen, D., Osgood, D., Zilberman, D., 2000. Publicly funded weatherdatabase benefits users statewide. Calif. Agric. May–June, 21–25.

44. Pereira, L.S., Cordery, I., Iacovides, I., 2012. Improved indicators of water use perfor-mance and productivity for sustainable water conservation and saving. Agric.Water Manage. 108, 39–51.

45. Perry, C., Steduto, P., Allen, R.G., Burt, C.M., 2009. Increasing productivity in irri-gated agriculture: agronomic constraints and hydrological realities. Agric. WaterManage. 96, 1517–1524.

46. Portoghese, I., D'Agostino, D., Giordano, R., Scardigno, A., Apollonio, C., Vurro, M.,2013. An integrated modelling tool to evaluate the acceptability of irrigationconstraint measures for groundwater protection. Environ. Modell. Softw. 16,90–103.

47. Pulwarty, P., Maia, R., 2013. Adaptation challenges in complex river basins aroundthe world. In: Eighth International Conference of EWRA, Water Resources Man-agement in an Interdisciplinary and Changing Context, 26–29 June, Porto,Portugal.

48. Rita, C., Capelo, O., 2005. Distribuic, ão em alta pressão vs. distribuic, ão embaixa pressão num bloco de rega do E. F. M. Alqueva (Monte Novo). In:Paper in I Congresso da Rega e Drenagem, Beja, Portugal, _Available inwww.cotr.pt/informacao/web2/Papers/38.pdf_.

49. Rosado, L., 2012. A cultura do milho na Fundac, ão Eugénio de Almeida. In: VIII Con-gresso Nacional do Milho, Lisboa.

50. Salvador, R., Martínez-Cob, A., Cavero, J., Playán, E., 2011. Seasonal on-farm irrigationperformance in the Ebro basin (Spain): crops and irrigation systems. Agric. WaterManage. 98, 577–587.

51. Santos, J.F., Gomes, J.C., Fialho, I., 2011. Contributo para a discussão dascondic, ões de sustentabilidade da component hidroagrícola do EFMA, Jor-nadas Técnicas APRH, Lisboa, LNEC, 12-15 Outubro, http://sir.dgadr.pt/conteudos/jornadas aph/apresentacoes/s2/32.pdf, Technical Days index athttp://sir.dgadr.pt/jornadasaprh

52. Silva, A.M., 2010. Alqueva: água baixa 70% para agricultores, Correo da Manhã,11 abril, http://www.cmjornal.xl.pt/detalhe/noticias/nacional/economia/agua-baixa-70-para-agricultores, http://boasnoticias.sapo.pt/mobile/noticias.php?id=771Sinclair,

53. T.R., Tanner, C.B., Bennett, J.M., 1984. Water-use efficiency in crop produc-tion. BioScience 34 (1), 36–40.

54. Steduto, P., 2007. On the conservative behaviour of biomass water productivity. Irrig.Sci. 25 (3), 189–207.

55. Tollefson, L.C., Wahab, M.N.J., 1994. Better Research-Extension-Farmer Interactioncan Improve the Impact of Irrigation Scheduling Techniques. World Bank, Wash-ington, DC.

56. Vergés, J.C., 2001. The Alqueva Irrigation Agreement (Study subcontracted to Mod-ern Consulter S.L. by Research Contractor WRc plc, for the Environment Agency,UK).

57. WBCSD, 2000. Eco-efficiency. In: Creating More Value with Less Impact.World Business Council for Sustainable Development, _http://www.wbcsd.org/web/publications/eco efficiency creating more value.pdf_.

58. WssTP, 2012. Irrigated Agriculture Water Saving Options in Irrigation, Lookingfor Efficient Techniques, Irrigation Management and Adapted CroppingPractices. Water Supply and Sanitation Technology Platform (WssTP),_http://www.wsstp.eu/files/WSSTPX0001/library/agriculture/SP%20Agriculture%20VI.pdf_.

59. WssTP, 2013. Water Reuse: Research and Technology Development Needs forWater Reuse. WssTP, Summary at _http://www.wsstp.eu/files/WSSTPX0001/Water%20Reuse%20Ex.Summary%20online.pdf_.

60. Zaccaria, D., Oueslati, I., Neale, C.M.U., Lamaddalena, N., Vurro, M., Pereira, L.S.,2010. Flexible delivery schedules to improve farm irrigation and reduce pres-sure on groundwater: a case study in southern Italy. Irrig. Sci. 28, 257–270.

61. Zaccaria, D., Lamaddalena, N., Neale, C.M.U., Merkley, G., 2013. Simulation ofpeak-demand hydrographs in pressurized irrigation delivery systems using adeterministic-stochastic combined model. Part II: Model applications. Irrig. Sci.31, 193–208.

4

CHAPTER

From Unplanned to Planned Agricultural Use: Making an Asset Out of Wastewater

Jules B[1,2], van Lier[1,2],Frans P. Huibers[3]

[1]*Faculty of Civil Engineering and Geosciences, Department of Water Management, Section Sanitary Engineering, Delft University of Technology, P.O. Box 5048, 2600 GA Delft, The Netherlands*

[2]*Department of Water and Sanitation, Unesco IHE Institute for Water Education, P.O. Box 3015, 2611 AX Delft, The Netherlands*

[3]*Irrigation and Water Engineering Group, Wageningen University, P.O. Box 47, 6700 AA Wageningen, the Netherlands*

ABSTRACT

Urban wastewater is increasingly used for agricultural production, particularly in those areas where access to fresh water resources is limiting. Depending on the prevailing institutional arrangements, this agricultural use is planned or unplanned. If planned, a general policy is to minimise health risks and environmental pollution, leading to an often centralised collection of the city's sewage, followed by primary, secondary and further treatment until the prevalent discharge effluent use standards are met. If society can afford them, advanced treatment technologies are installed, backed by a well functioning institutional infrastructure. In contrast, in less prosperous countries the agricultural use of wastewater is driven by the high needs for water and the absence of affordable fresh water resources along with insufficient financial means to construct treatment systems and distribution networks that comply with the official regulations. Combined with a poor institutional framework, this results in unplanned and unguided direct or indirect use of raw, partially treated or diluted wastewater. In an effort to mitigate health and environmental problems, we propose using a reverse water chain design approach, in which the ultimate fate of the water is the basis for the design of conveyance and treatment facilities.

Keywords: Design Developing countries Peri-urban agriculture Reverse water chain Urban water management Wastewater irrigation Wastewater treatment Water quality

INTRODUCTION

In many developing countries, water is becoming a limiting factor and wastewater is considered an alternative water resource for agricultural production. After all, a small town of 100.000 people produces sufficient wastewater to cultivate an area of up to 100 hectares, all year around, assuming a per capita urban water consumption of 80–100 l/day. The major constraint in irrigation with wastewater is the hazards that it poses to human health. Environmental issues also play a role, such as the fate of nutrients outside the irrigation season and the effects of micro-pollutants, although these problems are worse if all the wastewater is directly discharged.

In industrialised countries these constraints are addressed by implementing extensive treatment systems that eliminate the mentioned health and environmental hazards (e.g. Asano 1998), in line with a conventional urban water chain approach (Huibers and van Lier 2005), in which water is centrally distributed to houses and industrial complexes and the produced wastewater is subsequently collected by sewerage systems. The treatment plant is the targeted location for intercepting the water quality in order to meet regulations for discharge or effluent use.

By directly linking a treatment plant to an agricultural production site, the required treatment technology needs reconsideration, taking into account the effluent characteristics of actual concern in agricultural use (Martijn and Huibers 2001).

- ***COD/BOD*** levels can be more relaxed than the generally applied stringent values, particularly considering the positive effect of organic matter on soil structure in agricultural fields that are deprived of organic matter.
- ***Nutrients*** are resources that can beneficially be used by farmers and should not be removed. However, off-season or whenever nutrients are in excess, periodic nutrient removal could be a strategy in the treatment system (Kassab et al. 2007). Depending on the crops that are grown, the available nutrients are sufficient for crop cultivation,

reducing farmer's expenditure on artificial fertilizers (Boom et al. 2007). Balancing nutrients in the field to support crop requirements, while avoiding environmental pollution, gives important incentives to reconsider the wastewater treatment techniques used, as well as the irrigation water management system.

- ***Pathogens*** removal is considered an important step in the treatment process. However, the problem with pathogens can also be addressed at other levels in the water chain or in the food chain as is discussed in this paper.
- ***Solids*** could have a negative impact on irrigation, particularly in pressurized systems. A step to remove might be included in the wastewater treatment system, or alternatively in the irrigation network through additional filtration or settling steps, e.g. in storage basins.
- ***Heavy metals*** are generally not considered when treating domestic wastewater, as they are generally not in. However, this might be different in developing countries, certainly where wastewater origin is not always well defined. When coupled with agricultural use, care should be taken since heavy metals accumulate in the soil. A high degree of heavy metal removal is expected when an anaerobic treatment step is included, causing precipitation of (bivalent) metal-sulphides (van Lier 2008).
- ***Salinity*** of the effluent is not considered in conventional treatment, because salts cannot be removed by plain technologies, but it is of utmost importance if the effluent is used in irrigation. This parameter is determined by the sources of wastewater and the level of evaporation in the treatment process.
- If an irrigation system fully depends on effluent as a water source, operational storage of effluent is needed. This storage facility would be part of the overall development and should be designed such as to play a role in (post-) treatment.

THE CONVENTIONAL WATER CHAIN APPROACH

Figure 1 depicts a typical representation of a conventional urban water management chain, in which upstream water source and downstream effluent use are *de facto* considered as black boxes. Drinking water arrives at the households at the highest quality. From here, the used and contaminated water is collected in a sewerage system, where rainfall runoff and other pollutants are added. Before the collected sewage is discharged into the environment or used in agriculture, treatment is required to reach agreed standards. Scientific insights and public perception of potential risks that can occur as a result of using treated wastewater, lead to public demands to tighten the quality criteria, requesting efficiencies for wastewater treatment at the highest level.

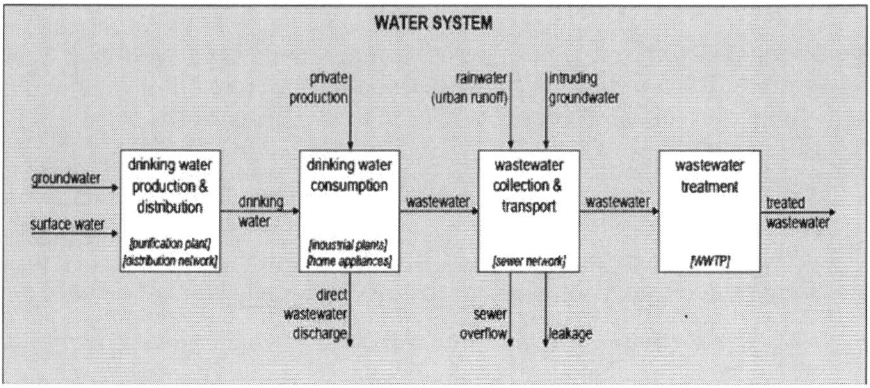

Figure 1. Schematic representation of the conventional urban water chain (Bots 2008)

In addition to numeric values obtained from environmental and human health studies, standards are also getting stricter as a result of emotional debates and legal liability of those producing the treated water. The ever-increasing required treatment efficiencies with regard to pathogens, micro-pollutants and nutrients in the arid areas of industrialised countries such as parts of the USA and Israel are illustrative in this respect (see Shuval et al. 1997; Rose 2007). In industrialised countries, more and more technologies are under development that could be used to achieve this goal. The costs for infrastructure, operation and maintenance, and institutional guidance are increasingly higher, yet covered by tax income or are invoiced to the specific polluter in the chain. Shuval *et al.* estimated that based on the USEPA/USAID guidelines, the costs of wastewater treatment to prevent a single case of Hepatitis A, would be approximately US$ 30 million. This, of course, by far exceeds the costs for curing a patient who has fallen ill. Moreover, Blumenthal and Peasy (2002) state that there is no epidemiological evidence proving the need to define the guidelines more strictly than the WHO values of 1989 (WHO 1989).

In less prosperous countries, civil and institutional infrastructure is poorly developed and sometimes largely lacking. Nonetheless, the need for the efficient use of water is huge, particularly in water stressed countries. Yet, the general concern about environmental and human health issues related to the use of effluents in irrigated agriculture is apparent in each country. Official governmental regulations often demand stringent wastewater treatment requirements, especially when the effluent is used for producing edible crops.

As financial means are by far insufficient to construct and maintain the required infrastructure, this creates a paralysed situation, which aggravates when the role of agriculture for city food production and for providing a livelihood for the urban and peri-urban farmers, remains unrecognised. The net result is an unofficial tolerance of the uncontrolled usage of polluted water as can be observed in many countries around the globe. The total agricultural area irrigated with partly or untreated wastewater exceeds 20 millions ha in over 50 countries (van der Hoek 2004).

As discussed, costs of distribution, collection, treatment and logistic management in a conventional urban water chain are extremely high and the required infrastructure needs a long period of time for development. The sewerage system in the Netherlands, with a population of 16 million, has a total length of 109,000 km, which is 7.5 m per person (Stichting Rioned 2006). The yearly maintenance costs of this sewerage system, including the treatment plants, is approximately € 150 (US$ 200) per capita. In addition to the resources required to construct the civil and electro-mechanical works, energy requirements for pumping and wastewater treatment are far beyond the budget of most developing countries. Accepting that the conventional approach makes no sense for many of the less prosperous countries, alternative strategies should be searched for which do not compromise on the environmental and human health constraints.

THE REVERSE WATER CHAIN APPROACH

In a reverse water chain approach, the quantitative and qualitative water requirements of the final users are taken as the starting point for designing the upstream civil and institutional infrastructure. As a consequence, the type and functionality of the wastewater treatment system is determined by the farmer's need rather than by governmental legislation for wastewater discharge.

Such a reverse approach leads to design criteria that differ from those in a conventional design:

- The required *effluent characteristics* (pathogens, nutrients, organics, suspended solids and toxicants) of the water to be used for irrigation are flexible and negotiable, as these depend, among other things, on the crops to be grown; the irrigation technique and water management to be applied; the further crop handling; as well as the acceptability and perception of farmers and consumers.

- The *location* of the treatment plant would be defined in relation to the availability of suitable agricultural field, assumed to be available in the immediate peri-urban areas. This would reduce costs of effluent transport.
- Optimizing the location of the treatment plant *vis a vis* agricultural fields would in many cases require a policy of *decentralization*. This leads to an overall reduction of the length and size of the sewerage system and would allow a better control of the wastewater inflow, excluding e.g. toxic industrial waste streams into the system.
- To solve the mismatch between effluent supply and irrigation water demand, *effluent storage* in the irrigation system is needed, which should also be considered for its role as post-treatment step.
- Knowing the benefits of effluent use determines to a large extent the *economic affordability* with respect to the required infrastructure and the acceptable costs per volume of treated sewage.
- Downstream productive use should lead to a policy of *separating waste streams* avoiding the mixing of black water with large amounts of relatively clean waters.

Using a reverse water chain approach would imply intensive discussions and negotiations between all stakeholders concerned. In such a process other possible interventions might be identified in addition to the wastewater treatment process, with the aim to reduce overall costs and to fully access the available resources, while reducing uncontrolled discharges into the environment. Subsequently, related costs can be spread across different stakeholders.

PATHOGEN HANDLING IN A REVERSE WATER CHAIN APPROACH

Contamination of agricultural products by pathogenic organisms is considered to be one of the major risks in effluent use. The generally accepted approach is to treat the wastewater up to levels at which the contamination potential is practically eliminated. Consequently, high-cost advanced wastewater treatment systems are required. In practice, such extensive pathogen removal efficiency is hardly ever realized in developing countries. However, by analyzing the route from the cultivation of the crop into the household, various levels of interception can be distinguished showing that pathogen removal and pathogenic contamination do not necessarily have to be addressed by advanced treatment alone (WHO 2006) but can also be made at the stages of crop handling, crop selection and irrigation.

Crop handling

Using clean irrigation water does not give a guarantee that a crop is free of pathogens when sold (Amoah et al. 2005), as also crop handling is a potential source of contamination. This is particularly true in less prosperous countries where numerous sources of infection are apparent. Safe handling requires knowledge transfer and education of farmers, field labourers, transporters and merchants as well as consumer.

Crop selection

Depending on the quality of the effluent to be used, the crops to be cultivated can be selected by degree of wastewater contact, in which the highest degree of purification is required for crops that are in full contact with the irrigation water and those that are eaten raw (WHO 2006).

Irrigation water management

Pathogen-rich wastewater should be handled with care to avoid direct contact by farm labourers and people in the vicinity. Use of sprinkler irrigation is discouraged as pathogen loaded droplets (aerosols) may drift over large distances. On the other hand, irrigation techniques such as drip irrigation limit the contact between the wastewater and the irrigators and crops.

The field could also be used for disinfection, for example by fixing a required minimum period between the last irrigation and the crop harvesting. In this way UV-sunlight and field-drying are used as means to reduce the pathogen levels on the crops (Pescod 1992), although this does not apply to the more persistent cysts and the helminths ova or worm eggs.

Irrigation scheme

After treatment, the water is distributed to an irrigation network. For operational reasons, in order to regulate peak demands as well as fluctuations in effluent flow, the irrigation network should be preceded by a storage facility. This allows for an optimal use of the effluent and nutrients (Darwisch et al. 2007). Such a reservoir could be combined with pathogen removal, reducing the costs for the overall layout. Disinfection taking place in such reservoirs include settling of helminth ova and natural decay of pathogenic organisms. In addition, low amounts of chemicals or other disinfection methods could be used.

Access to fresh water in addition to the effluent creates the possibility for blending different water qualities, an excellent water management technique not only to balance the water demand-supply ratio, but also to lower concentrations of pathogens, along with nutrients, and salts.

Treatment technology

Disinfection of wastewater could be part of the applied treatment technology. Considering the negative impact of chlorine; the formation of persistent carcinogenic chlorinated organic compounds, other methods such as ultraviolet (UV) radiation, titanium oxide (TiO_2) and membranes should be used in order to reduce the pathogen load. Particularly when treatment plants do not deliver high quality effluents, chlorination deteriorates the quality rather than improving it. Re-growth of indicator organism in the effluents has also been observed (WERSC, University of Jordan, unpublished).

It must be noted that advanced pathogen removal, is generally a very costly part of the treatment scheme. A more cost-effective approach is to integrate field-pathogen removal as discussed in the crop handling subsection with removal of persistent cysts and helminth ova in the treatment plant. Helminth ova are relatively big and can be removed from the water using techniques such as settling and filtration using packed beds or upflow anaerobic sludge blanket systems (Von Sperling and de Lemos Chernicharo 2005; Jimenez 2007).

Sanitation system

The general concept adopted in industrialised countries is flush-through toilets, which are connected to the sewerage network. In this way, the concentrated hazardous waste (human faeces) is diluted by a factor of 100–200 or more, creating an immense pool of diluted infectious wastewater. This will get worse if urban runoff is also conveyed with the same system. Instead of combined sewage systems, the introduction of source separated sewerage, where the black (toilet) water is separated from the rest, will result in grey wastewater which is relatively easy to treat and which contains several million fewer pathogens. Particularly in new housing estates, hotels, hospitals, and recreational areas, source separating sewer systems are relatively easy to construct, yet hardly implemented.

The previous analysis illustrates that by following an integrated approach, constraints related to the use of sewage effluents in irrigated agriculture can be dealt with at various locations in the water chain. Such an approach can also be applied to other compounds of concern like nutrients, micro-pollutants and salts.

WASTEWATER TREATMENT FOR EFFLUENT USE IN IRRIGATED AGRICULTURE

Following a reverse water chain approach, the selection of the most appropriate treatment method, with the aim to use the effluent as irrigation water, would follow other criteria than the conventional approach. The following criteria are of importance, the weight of which is dependent on the local socio-economic situation:

Cost effectiveness

The availability of cost-effective treatment systems will increase the accessibility of sewage as a water and nutrient resource and will reduce the need for farmers to use raw sewage in agriculture. Obviously, the costs for the required investments are a community burden and can never be absorbed by the poor farmers that now use the raw sewage. However, by implementing an integrated approach, a shared cost concept could lead to a feasible solution. Alternatively, a private company could take the delegated responsibility for wastewater treatment, charging treatment fees to the citizens and water supply fees to the farmers. Long-term depreciation periods, soft loans and government regulations are tools for preventing excessive fees for either citizens or farmers. Obviously, implementation and control of sewage treatment requires a well-developed institutional framework, irrespective of the type of technology that is applied.

Compactness

An important advantage of compact systems is the possibility to construct the treatment in the vicinity of the populated area, preventing the need for expensive conveyance networks and improving control over the origin of wastewater flows. Surface systems such as pond systems, constructed wetlands, and slow sand filtration require large areas of land and thus are constructed at some distance from the city, where land is cheaper. This distance is also required to avoid odour problems in the urban area. From an agricultural point of view, it is obvious that direct evaporation from pond systems, which could count for 15% or more of design inflow in hot climates, and a concomitant increase of the salt concentration leads to poorer quality water. On the other hand, the surface systems do have their advantages as cheap post-treatment, or as part of the effluent storage capacity, so should be considered if location permits.

Robustness

A treatment system must be reliable for all circumstances. Irregularities in flow and load should not disturb the processes of the selected treatment system. In addition, the dependence on infrastructure investments, such as electricity supply, should be as low as possible. Also, the costs of electricity could be prohibitive, because in developing countries the availability of electricity supply is not always guaranteed 24 h day^{-1}. The same is true for regions in conflict.

Flexibility

It is not very likely that at present developing countries will be in the position to copy "Western standards" and treatment capacity when developing and implementing their local policy. Probably a similar approach will, and needs to be, followed as we have seen in the industrialised world. This means, a gradual increase of discharge standards allowing the implementation of basic treatment first, followed by a gradual extension of the treatment plant to more stringent legislation. It is well known that the removal of 80% of the pollution load can be obtained with limited costs, while the costs for the removal of the last 20% does not merit its purpose if in the same region raw sewage is discharged owing to limited availability of funds. But the chosen system must allow future extension to more stringent legislation.

Acceptability

The technology to be applied, with clear reference to affordability, must be accepted by the various stakeholders involved, including the executing institutions, responsible engineers, local community and farmers, as this will guarantee the sustainability of the investment. Often a discrepancy is observed between the wishes of responsible institutions and community members, and the capabilities of the local community, particularly in donor projects where technology-push is a consequence of a supply driven set-up of a project.

Decentralisation

As the generally applied "Western" concept is based on centralised services, huge infrastructure investments are required. For developing countries, the construction of extensive sewerage networks is absolutely out of the question, forcing a change in conceptual thinking. In a decentralised approach, a city, region, or country can be served with sanitation and treatment step by step, without the need to firstly accomplish a regional infrastructure investment. In this sense, decentralisation will immediately increase the accessibility of sewage as an alternative water and nutrients resource, even within the urban context. Moreover, the decentralised approach provides a means to exclude heavily polluted wastewaters from industrial estates from the domestic wastewaters collected. Industrial discharges may severely limit the applicability of treated wastewaters in the agricultural field owing to the presence of heavy metals, toxic compounds, persistent organics, salts, among other constituents.

The decentralised set-up, using satellite gravity sewers in a single catchments area, can be constructed using low cost sewerage systems. For instance, smallbore sewers connected to (available and improved) septic tanks, now considered to be constructed in Bonaire (Haverkamp 2009), offer a cost-effective solution for primary wastewater treatment combined with conveyance to a collection point. A compact decentralised post-treatment, such as a trickling filter, may then upgrade the effluent quality until the required standards for (restricted) irrigation. From the water distribution point of view, decentralisation has the advantage of limited conveyance of the treated water as well. A small collection area also implies a small distribution system after treatment. The latter can be more easily combined with peri-urban farming systems.

Location

In order to avoid the need for large conveyance networks for the treated water, the treatment systems should, ideally, be constructed closer to agricultural fields where the treated water will be used. In the conventional civil engineering point of view, treatment systems are generally located at the lowest point, which, in coastal areas, is near to the sea (Australia!) where there are no agricultural activities. This is particularly the case in a centralised set-up of the master-management plan e.g. in the case of Tunisia (Chenini et al. 2003).

Management plan

For the sake of sustainable development, a master-management plan must underlie the individual projects. In considering a decentralised set-up, such a management plan might look more complex, but could also be developed stepwise. An interesting reference is the city of Recife, Brazil that several years ago adopted the decentralised approach, including decentralised wastewater treatment as the most feasible alternative in their specific situation (Florencio and Kato 2001).

DISCUSSION

A reverse water chain approach is a new concept of wastewater management and offers interesting perspectives for economising the urban water chain. In eThekwini, South-Africa, such approach is being tried out in a collaborative effort between municipality, its agricultural management unit, consultants, farmers and research institutions aiming to develop an alternative for the unsuccessful top-down approach applied before (pers. comm. Chris Buckley and Wolf Raber). In contrast to the generally applied approach, the ultimate users, *i.e.* peri-urban farmers, set the criteria for designing the urban water flows. Separation of black toilet waters from the other wastewater streams and decentralised collection, to avoid the inclusion of industrial wastewater, will more easily facilitate the use of the treated water in agriculture. Moreover, treatment systems would be located nearer to the agricultural fields and can be simplified if the fields are incorporated in the overall treatment scheme. In return, farmers can benefit from the available water and nutrients to produce the crops. Whether or not this is acceptable depends on the local situation and the socio-economic conditions of the farmers.

OPEN ACCESS

REFERENCES

1. Amoah P, Drechsel P, Abaidoo RC (2005) Irrigated urban vegetable production in Ghana: sources of pathogen contamination and health risk elimination. Irrig Drain 54:S49–S62
2. Asano T (ed) (1998) Wastewater reclamation and reuse. Water quality management library, Vol. 10. Technomic Publishing Company, Inc, Lancaster
3. Blumenthal UJ, Peasy A (2002) Critical review of epidemiological evidence of the health effects of wastewater and excreta use in agriculture. Unpublished document prepared for WHO, Geneva (available on request), 40 pp
4. Boom S, Huibers FP, van Lier JB (2007) Wastewater irrigation in Jordan: a mismatch in macro nutrient provision. Water Practice and Technology, 3(2):
5. Bots PWG (2008) Benchmarking in Dutch urban water management: an assessment. Adaptive and Integrated Water Management: 277–300
6. Chenini F, Huibers FP, Agodzo SK, Van Lier JB, Duran A (2003) Use of wastewater in irrigated agriculture. Country studies from Bolivia, Ghana and Tunisia. Vol. 3: Tunisia. Wageningen: Wageningen Universiteit, IWE (W4F-Wastewater), pp 52
7. Darwisch MR, Sharara M, Sidahmed M, Haidar M (2007) The impact of a storage facility on optimality conditions of wastewater reuse in land application: a case study in Lebanon. Resour Conserv Recycl 51:175–189
8. Florencio L, Kato MT (2001) Perspectives of anaerobic treatment for domestic sewage in Recife Metropolitan Region. In: van Lier JB, Lexmond M (eds), Anaerobic digestion for sustainable development. Proc. Int. Farewell Seminar of Prof. Gatze Lettinga, 29–30 March 2001, Wageningen, The Netherlands
9. Haverkamp Y (2009) Water and wastewater in the Caribbean. An overview of the current and future water and wastewater situation on the Caribbean islands. Wageningen University, unpublished MSc-thesis
10. van der Hoek W (2004) A framework for a global assessment of the extent of wastewater irrigation in developing countries: health effects and technical solutions. Technical Paper No. 51. World Bank, Washington DC, USA

11. Huibers FP, van Lier JB (2005) Use of wastewater in agriculture: the water chain approach. Irrig Drain 54:S3–S9

12. Jimenez B (2007) Helminth ova removal from wastewater for agriculture and aquaculture reuse. Water Sci Technol 55(1–2):485–493

13. Kassab GH, Koetse E, van Lier JB (2007) Integrating methanogenesis and denitrification in EGSB reactors for adjusting effluent nitrogen level. In: Proc. of 6th IWA Specialist Conference on Wastewater reclamation and reuse for sustainability, Antwerp, Belgium, October 9–12, 2007

14. Martijn E-J, Huibers FP (2001) Use of treated wastewater in irrigated agriculture. Treated wastewater characteristics and implications. CORETECH Working document WP4-1: 34 pp. ISBN: 90-6754-649-6

15. Pescod MB (1992) Wastewater treatment and use in agriculture. Irrigation and Drainage Paper 47, Food and Agriculture Organization, Rome

16. Rose JB (2007) Water reclamation, reuse and public health. Water Sci Technol 55(1–2):275–282

17. Shuval H, Lampert Y, Fatal B (1997) Development of a risk assessment approach for evaluating wastewater reuse standards for agriculture. Water Sci Technol 35(11–12):15–20

18. Stichting Rioned (2006) Riool in Cijfers, 2005–2006. Periodic report (in Dutch). Available on internet via: http://www.riool.net/riool/binary/retrieveFile?instanceid=20&itemid=61&style=default

19. Van Lier JB (2008) High-rate anaerobic wastewater treatment: diversifying from end-of-the-pipe treatment to resource oriented conversion techniques. Water Sci Technol 57(8):1137–1148

20. Von Sperling M, de Lemos Chernicharo CA (2005) Biological wastewater treatment in warm climate regions. IWA, London. ISBN 1 84339 002 7

21. WHO (1989) Health guidelines for the use of wastewater in agriculture and aquaculture. Technical report series 778. World Health Organisation, Genève, Switzerland

22. WHO (2006) Guidelines for the safe use of wastewater, excreta and grey water, vol. 2. Wastewater use in agriculture. World Health Organisation, Genève, p 213

5

CHAPTER

Matric Potential-Based Irrigation Management of Field-Grown Strawberry: Effects on Yield and Water Use Efficiency

Guillaume Létourneau*, J. Caron, L. Anderson, J. Cormier

Department of Soil and Agrifood Engineering, Laval University, Pavillon de l'Envirotron, 2480 boul. Hochelaga, Québec G1 V 0A6, Canada

ABSTRACT

Effective and adapted criteria for irrigation scheduling are required to improve yield and water use efficiency (WUE) and reduce the environmental impacts associated with water and nutrients losses by runoff and leaching. In this study, field-scale experiments were conducted at four commercial strawberry production sites with contrasting soil and climatic conditions. Within each site, the influence of different soil matric potential-based irrigation thresholds (IT) on yield and WUE was evaluated. Matric potential-based irrigation management was also compared with common irrigation practices used by producers in each site's respective areas. At Site 1 (silty clay loam; humid continental (Dfb) climate), an IT of −15 kPa improved yields by 6.2% without any additional use of water relative to common irrigation practices. At Site 2, with similar soil and climatic conditions, the irrigation treatments did not affect yield and the matric potential-based management decreased WUE relative to common practices. However, the results suggested that maintaining the soil matric potential lower than −9 kPa could induce stressing conditions for the plants. At Site 3 (sandy loam; Mediterranean (Cs) climate), the best yield and WUE were obtained with an IT of −8 kPa and suggested that WUE could be further improved by implementing high-frequency irrigation. At Site 4 (clay loam; Mediterranean (Cs) climate), results suggested that an IT between −10 and −15 kPa could optimize yield and WUE, and matric potential-based irrigation considerably reduced leaching under the root zone relative to common practices. Considering the results from all sites, an IT of −10 kPa appears to be adequate as a starting point for further optimizing irrigation under most field conditions.

Keywords: Strawberry; Irrigation management; Water use efficiency; Soil matric potential;Tensiometer

1. INTRODUCTION

Irrigation management is of primary importance for the profitability and sustainability of field strawberry production because it affects yield, water use efficiency and diffuse pollution of ground and surface water. These issues are of great importance in some areas, such as coastal California, Spain and Australia, where water availability is a growing concern. In recent decades, the increased implementation of more efficient water management practices, mainly subsurface drip irrigation (SDI) and the use of plastic mulch, has greatly improved WUE in commercial strawberry production. However, questions remain regarding irrigation management criteria that could improve WUE and reduce the environmental risks of SDI. Currently, the most common criteria are based on climatic water balance (evapotranspiration), plant physiological properties, soil water status measurements or a combination of these factors.

Many studies have shown that evapotranspiration (ET)-based irrigation management could be efficient for strawberry production (Hanson and Bendixon, 2004, Krüger et al., 1999 and Yuan, 2004). However, this method is also criticized for its inability to account for rapid changes in climatic conditions and because it generally does not account for differences in the water requirements of different strawberry cultivars (Giné Bordonaba and Terry, 2010, Klamkowski and Treder, 2008 and Krüger et al., 1999). The availability of locally determined crop coefficients that account for the wetting patterns resulting from the combined effects of SDI system configuration and soil type can also be problematic. Additionally, management based only on ET cannot be used to assess whether the applied irrigation water is lost beneath the root zone due to percolation (Simonne et al., 2012).

Numerous studies have proven that physiological measurements on plants could be used to measure water stress and its effect on plant performance. However, plant-based irrigation scheduling is still limited by many theoretical and practical difficulties, most of which are discussed in a comprehensive review by Jones (2004). In strawberry production, leaf temperature measurements with infra-red thermometers were identified as a potential tool for irrigation management because they allow for the detection of severe stresses that affect yield and biomass production (Peñuelas et al., 1992). Photosynthesis, stomatal conductance, and leaf water potential measurements have also been successfully used to detect hydric stress and to understand and differentiate stress adaptation mechanisms among strawberry cultivars. Such measurements have shown great potential to be used for breeding drought-resistant cultivars (Blanke and Cooke, 2004, Klamkowski and Treder, 2006, Klamkowski and Treder, 2008 and Savé et al., 1993). Canopy spectral reflectance or transmittance measurements also present potential for irrigation management because they are correlated with strawberry fruit yield. However, a clear interpretation of their effects still requires topography, soil water content and climatic information (Hoppula and Salo, 2007). Despite the relevance of understanding the physiological parameters tackled in the studies cited above, they do not currently allow for the definition of a criterion for irrigation scheduling or for assessing the appropriate volume of water that should be applied (Jones, 2004).

Soil water status measurement methods, whether based on water content or soil matric potential, have been well documented (Topp and Ferré, 2002 and Young and Sisson, 2002) and have frequently been used for irrigation scheduling. In this study, soil matric potential (ψ) measurements were favored over soil water content measurements because of their ease of use in various field conditions (no calibration required for soil types or salinity levels). The soil matric potential is also directly related to the soil's hydraulic conductivity, which is linked to the soil's capacity to supply water at the rate required by plants (Rekika et al., 2014). Because potential gradients are the leading force responsible for water movement in soils, ψ measurements can also be used to infer the direction and magnitude of soil water fluxes. This information can be used to prevent leaching under the root zone (Krüger et al., 1999). ψ-based irrigation management has been successfully used to improve the yields of many agricultural crops and to evaluate the impacts of irrigation practices on water and fertilizer leaching (Pelletier et al., 2013,Périard et al., 2012 and Shock and Wang, 2011).

Many studies have evaluated the impacts of ψ-based irrigation management on strawberry yield and WUE. However, their results differ with regard to which ψ value should be used as an irrigation threshold (IT). Similarly, little information is generally provided regarding the optimal duration of each irrigation event. From a field experiment in a sandy loam presenting a significant proportion of coarse particles ($\varnothing$ > 2 mm),Bergeron (2010) noted an increase in WUE that did not affect yield when the IT was −18 kPa. For strawberry grown in a sandy loam under a high tunnel, greater yields were obtained when an IT of −10 kPa was used than when a drier regime of −70 kPa was used (Guimerà et al., 1995 and Peñuelas et al., 1992). Under similar soil and production conditions, significant yield and fruit quality decreases were observed for ITs of −30, −50 and −70 kPa relative to an IT of −10 kPa, and the highest WUE was obtained when using an IT of −50 kPa. In the same study, the yields from all of the treatments were 150–250% greater than the average yields from growers in the area (Serrano et al., 1992). Based on field experiments, a similar trend was observed in sand with an optimal yield at an IT of −15 kPa and a better WUE at −30 kPa (Hoppula and Salo, 2007). In addition, irrigation regimes were shown to affect fruit quality during greenhouse production. Deficit irrigation was shown to increase the sugar/acid ratios, antioxidant capacity and total phenolic contents of strawberries (Giné Bordonaba and Terry, 2010 and Terry et al., 2007).

Nonetheless, not all studies support the previously mentioned benefits of ψ-based irrigation management. After conducting field experiments during three production seasons, Krüger et al. (1999) concluded that ET-based management showed a greater potential for improving yield and WUE, mainly due to economical and practical reasons. However, only one ψ-based irrigation treatment was tested, and irrigation scheduling was based on bi-weekly measurements. Problems with regards to probes locations in the field were also reported. In a bell pepper production experiment, it was shown that the soil matric potential presented large spatial and temporal variability and that several tensiometers were necessary for obtaining representative values for a given field. (Hendrickx and Wierenga, 1990 and Hendrickx et al., 1994).

Some of the previously mentioned studies were conducted in greenhouses or tunnels and others in fields. Additionally, the tensiometers installation depths, matric potential measurement frequencies and treatment application procedures varied between studies. Considering this, the optimal IT for given soil and climatic conditions is not obvious. The aim of this study is to determine optimal matric potential based IT for field strawberry production with regards to yield and WUE for a variety of soil type and climatic conditions.

2. MATERIAL AND METHODS

2.1. Site descriptions

From 2010 to 2013, field-scale strawberry production experiments were conducted at four commercial sites that were chosen to cover a wide range of soil properties and climatic conditions. The site locations, soil types, climatic conditions and cultural practices are provided in Table 1. Generally, the experimental sites can be classified into two main groups. Group A includes Sites 1 and 2, which are located near Quebec City (Qc, Can) in an area with a humid continental (Dfb) climate where the growing season is short and generally rainy. The historical mean water budget (Precipitation – ETc) of the area is positive during the growing season, but the use of polyethylene mulches make supplemental irrigation necessary, especially in July and August. At these sites, irrigation water is provided by private reservoirs that are filled by snowmelt and rainwater. Thus, the irrigation water is free and usually available in sufficient volumes throughout the growing season. The day neutral "Seascape" cultivar was planted in mid-May and harvested from mid-July to early October, with a production peak in September. Group A sites were located near each other and their soils were very similar. However, Site 1 was a former pasture where strawberries were grown for the first time, and Site 2 had been cultivated in rotation with strawberries, sweet potatoes and oats for many years. Thus, the soil from Site 1 was expected to have hydraulic properties and nutrient element contents that were more favorable for plant growth than Site 2.

Table 1. General descriptions of the experimental sites.

Site	Location (lat,long)	Growing season	Reference period	Soil type	Climate	Rainfall[a] (mm)	Cultural Practices	Irrigation system specifications
1	Quebec city, Qc, CAN (46.90°,–70.94°)	May–October 2011	August–October 2011	Silty clay loam	humid continental (Dfb)	280	2 plant rows per bed; 1 drip line per bed	lateral : 1.6 cm lateral depth: 3 cm emitter spacing: 20 cm emitter flow rate: 0.5 ph
2	Quebec city, Qc, CAN (46.92°,–70.96°)	May–October 2011	August–October 2011	Clay loam	humid continental (Dfb)	280	2 plant rows per bed; 1 drip line per bed	lateral : 1.6 cm lateral depth: 3 cm emitter spacing: 20 cm emitter flow rate: 0.5 lph
3	Oxnard, CA, USA (34.15°,–119.15°)	October 2011–June 2012	February–June 2012	Sandy loam	Mediterranean (Cs)	97	4 plant rows per bed; 3 drip lines per bed; sprinklers in early stages	lateral : 1.6 cm lateral depth: 5 cm emitter spacing: 20 cm emitter flow rate: 0.34 lph
4	Watsonville, CA, USA (36.89°,–122.67°)	November 2012–October 2013	April–October 2013	Clay loam	Mediterranean (Cs)	24	2 plant rows per bed; 2 drip lines per bed; sprinklers in early stages	lateral : 1.6 cm lateral depth: soil surface emitter spacing: 20 cm emitter flow rate: 0.5 lph

[a] Data for the reference period only.

Group B includes Sites 3 and 4, which are located near Oxnard and Watsonville in the southern and northern parts of California (USA), respectively. Both areas have Mediterranean or dry subtropical (Cs) climates. At Site 3, a commercial short-day cultivar was planted in October and produced from February to June. At Site 4, locally developed commercial day neutral cultivars were planted in November, were slowly established during the winter, and produced berries from April to mid-October. Both of these sites were near the Pacific Ocean and were characterized by foggy mornings and clear afternoons. Furthermore, most of the precipitation in this region occurred between December and mid-March. Consequently, most of the production period was rain free.

2.2. Experimental setup

At each site, a typical management zone (i.e. a zone that was irrigated independently from the rest of the farm) was selected as the experimental area. That zone was divided in 3–7 blocs where two to three IT treatments based on ψ measurements were assigned in a randomized complete bloc design (Table 2). At Sites 1, 2 and 4, a control treatment consisting of the common irrigation management techniques used by local producers was also included in the experimental setup. In all cases, the supply line(s) for the drip irrigation laterals was(were) modified to allow for the independent management of each treatment. A typical setup from this study is schematically represented in Fig. 1. More site-specific details regarding the experimental setup are provided in Table 2.

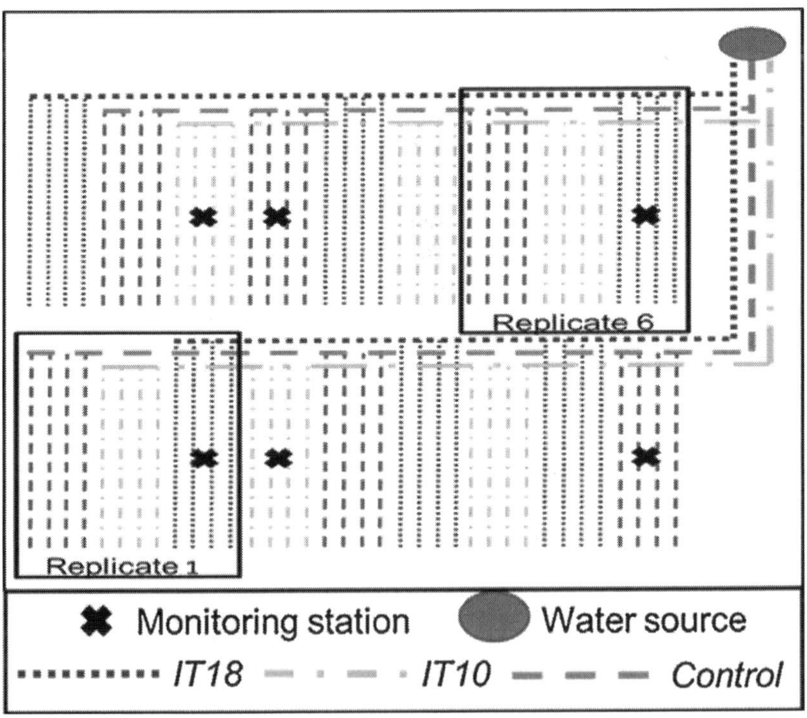

Figure 1. Typical setup of the drip irrigation system at the experimental sites (Site 1 in this example).Thicker lines represent principal and secondary conduits and thin lines represent drip laterals.

Table 2. Plot size, experimental design, yield measurement specificities and applied treatments on the experimental sites.

Site	Total site area (ha)	Bed length/width (m)	Beds per plot	Design	Harvest measurement frequency	Harvest measurement area	Plants per sub-plot	Applied treatments
1	1.4	142/0.9	4	RCBD 6 replicates	Bi-weekly	Total plot area	NA	*IT10*: −10 kPa *IT18*: −18 kPa *Control*: variable IT
2	1.8	160/0.9	4	RCBD 7 replicates	Bi-weekly	Total plot area	NA	*IT10*: −10 kPa *IT18*: −18 kPa *Control*: variable IT
3	0.7	130/1.22	3	RCBD 3 replicates	Weekly	Sub-plots only	24	*IT8*: −8 kPa *IT11*: −11 kPa *IT13*: −13 kPa
4	0.28	32/0.8	2	RCBD 4 replicates	Weekly	Sub-plots only	20	*IT10*: −10 kPa *IT17*: −17 kPa *IT25*: −25 kPa *Control*: variable IT

2.3. Irrigation system specifications

At all sites, the types of drip laterals used (diameter, emitter spacing, flow rate) and their configurations (nb of laterals per bed, depth of installation) were determined by the producer and were the same as those used on the rest of their farms. Site-specific details are provided in Table 1. For Group A sites, only drip irrigation was used for all growing stages. For Group B sites, overhead sprinkler irrigation was used during plant establishment. During that period, experimental plots were irrigated simultaneously according to the producers' usual practices. Treatment application began after the plant establishment stage, when only SDI was used to provide water.

2.4. Soil matric potential measurements

The soil matric potential was measured similarly at all the experimental sites. For all treatments, a ψ monitoring station was installed at mid-length in one (Group B) or two (Group A) replicates. Each monitoring station consisted of two tensiometers buried at two depths. The tensiometers were connected to a wireless transmitter (Hortau, St-Romuald, QC, CAN) to allow for real-time monitoring of ψ. A shallow probe was installed at a depth of 15 cm, where most strawberry roots are found (Bergeron, 2010). Horizontally, this shallow probe was located within a radius of 5–10 cm from a drip emitter. The aim was to place the probe in the active root zone and on the outer edge of the wetted bulb from the emitter. This position was used because it allowed monitoring the effects of root uptake and water application on ψ. A deep probe was installed at a depth of 30 cm, or 3 cm above any impermeable rock layer located at a shallower depth. Rock layers were present at three stations at Site 1 and two stations at Site 2 (out of 6 stations in both cases). Horizontally, the deep probe was located directly under a drip emitter to monitor the matric potential under the root zone (>90% of the roots are usually located within the top 30 cm), which is the area most likely affected by excessive irrigation.

2.5. Spatial variability of the matric potential

At Site 3, two manual reading tensiometers were installed at a depth of 15 cm in all of the experimental plots in early May. Thus, 6 probes were randomly distributed across the experimental area for each treatment. For each treatment, 6 measurement runs were performed from May 15 to June 15 to compare the ψ values obtained from the manual tensiometers with those obtained from the shallow probes of the monitoring stations. For each treatment and measurement run, Student's *t*-test with a 0.05 significance level was performed to test the null hypothesis that the average value from the manual probes did not differ from the value measured at the monitoring station.

2.6. Application of the irrigation treatments

The application of the experimental treatment was the same at all the experimental sites. For the IT treatments, irrigation was simultaneously triggered in all replicates of a treatment when a predetermined IT was reached by the shallow probe(s) from the monitoring station(s) associated with the treatment. When two monitoring stations were associated with a single treatment (Group A sites), irrigation was triggered when the average of the two shallow probes reached the IT. The ITs were chosen as a function of soil type or, in the case of Sites 1 and 2, previous experiments conducted on similar soils (Bergeron, 2010) and are presented in Table 2. The Control treatments consisted of the common irrigation management used by local producers. Irrigation of the *Control* plots was controlled by each site's producer without interference from the experimenters, but*Control* plots were still equipped with monitoring stations. For Group A sites, the common irrigation management method consisted of daily irrigation planning based on basic weather reports and irrigation logs from previous days. For Group B sites, the common practice was to apply approximately 100% of the historical average ETc via two or three weekly irrigation events on predefined days.

2.7. Adjustment of irrigation duration

For the IT treatments, the duration of each irrigation event was initially the same as that used in the *Control* treatments. For the *Control* treatments, the irrigation duration remained constant throughout the growing season. For the IT treatments, the irrigation duration was adjusted as follows. Shortly after planting, while the root uptake in the deeper parts of the bed was null, the ψ value at which the risk of rapid gravitational leaching under the bed was considered minimal was estimated using data from the deep probes of the monitoring stations. A typical example of changes in ψ at a depth of 30 cm following an early season irrigation is presented in Fig. 2. It shows that ψ at a depth of 30 cm sharply increases as the water content approaches saturation within minutes (Group A) or hours (Group B) after an infiltration (in this case irrigation) event. Then, if rapid gravitational leaching occurs, ψ rapidly (within 3 to 4 h) returns to a lower value. Next, a transition period between gravitational and capillary water flows usually occurs. Subsequently, after 16–48 h (depending on the site), water movements or changes in ψ are much slower and mainly result from capillary water redistribution. For the remainder of this study, ψ at the beginning of the capillary redistribution phase (Fig. 2) is referred to as the «bed capacity». This concept is similar to that of field capacity but more adapted to drip-irrigated raised beds, where the wetted bulb has a finite size (similarly to infiltration from an infiltrometer (Reynolds and Elrick, 1990)) and is not necessarily in hydraulic contact with the subsoil and water table. Drip irrigated raised beds generally reach their own equilibrium at ψ values that are much higher than those expected in a standard flat soil (−33 kPa). In this study, it was assumed that the risk of gravitational leaching was minimal at ψ values below the bed capacity. Using a trial and error process that generally required 1–2 weeks depending on the irrigation frequency, the irrigation duration was adjusted so that the bed capacity was not exceeded following irrigation unless leaching was desired to control the salinity in the root zone. Monitoring of the ψ data following irrigation was continued throughout the season, and the irrigation duration was adjusted if necessary. Because the bed capacities observed at all of the sites varied between −2 and −3 kPa, a target post-irrigation value of −3 kPa was used for all of the experimental sites considered in this study.

2.8. Cultural operations

All other cultural operations (pest, disease and weed management, flower and runner pruning and general maintenance) were executed similarly in all plots by the producers. Fertilizers were applied to the crops through injection into the drip irrigation system. Their type and quantity were controlled by the producer and were identical in all treatments. At Sites 1 and 2, the producer applied fertilizers to all plots simultaneously according to his usual fertilization schedule, regardless of the matric potential measured in the plots. At Sites 3 and 4, the *Control* treatments were fertilized regardless of the matric potential. However, for the IT treatments, fertilizers were only applied when the IT was reached. Thus, the timing of their application sometimes differed from 1 to 5 days among the treatments.

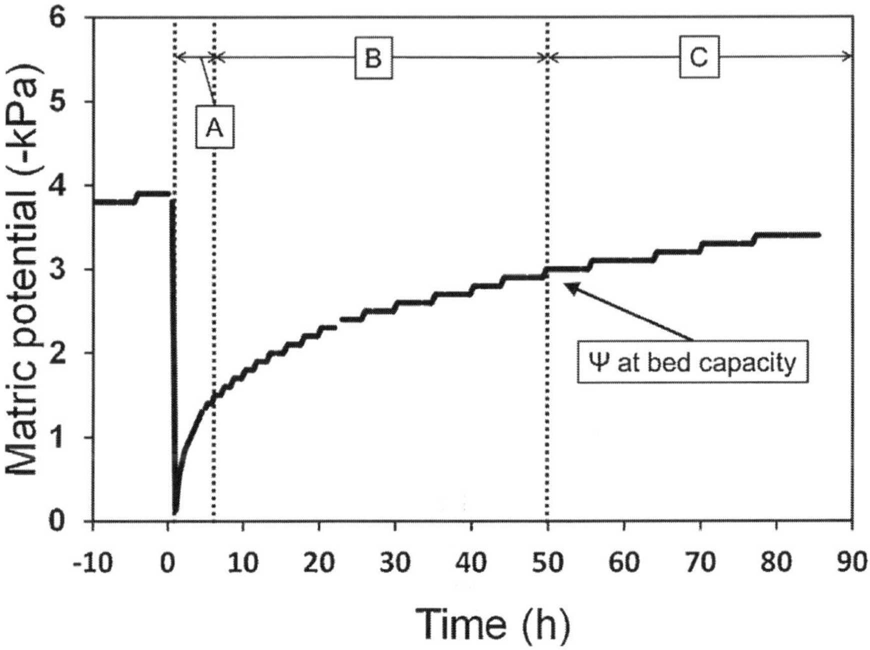

Figure 2. Example of changes in the matric potential under the root zone (30cm depth) following an irrigation event. Irrigation initiation occurred at $t = 0$ h. A is the gravitational flow period, B is the transient flow period and C is the capillary redistribution period.

2.9. Reference period for measurements and data analysis

For evident reasons, yield and fruit quality data were available only during the harvest period while water use, soil matric potential and climatic data were available for the entire growing season. In order to compare the influence of the experimental treatments on all variables of interest in this study, site-specific reference periods were established depending on data availability (Table 1). Consequently, all of the results presented in the remainder of this study regarding yield, fruit quality, applied water volumes, irrigation frequency and duration, achieved ITs and average ψ values (Table 3) only consider data obtained during that period. For Group A sites, the reference period was from August 1st to October 1st,

Table 3. Average thresholds, marketable yield, fruit quality and water use efficiency results from the different experimental sites, where ψ is the soil matric potential and WUE is the water use efficiency.

Site	Intended IT (treatment) (−kPa)	Avg achieved IT (−kPa)	Avg ψ 15 cm (−kPa)	Avg ψ 30 cm (−kPa)	Marketable yield[a] ($kg\,ha^{-1}$)	ETc (mm)	Applied irrig. water (mm)	Nb of irrig	Avg irrig. duration (min)	Irrig. freq (d)	WUE ($kg_{fruit}\,m^{-3}{}_{water}$)	Fruit size (g per fruit)	Sugar content (Brix index) (%)
1	*IT10*	13.3 ± 7.5	7.3	6.4	10687.9 ± 869.6 a	135.2	29	19	41	2.7	36.3	15.8 ± 0.5 a	8.8 ± 0.4 a
	IT18	14.9 ± 4.9	6.1	7.9	10518.3 ± 510.9 a		13	9	44	5.8	84.4	15.3 ± 0.7 a	8.8 ± 0.2 a
	Control	24.7 ± 6.8	10	7.2	**9981.1 ± 1151.6 b**		13	6	60	8.7	79.8	15.3 ± 0.7 a	8.8 ± 0.2 a
2	*IT10*	12.9 ± 8.6	5.0	4.5	12305.9 ± 1623.4 a	135.2	37	20	40	2.9	32.9	14.2 ± 0.8 a	8.2 ± 0.3 a
	IT18	18.0 ± 7.5	6.8	5.0	11492.9 ± 2087.2 a		29	15	42	3.8	41.5	14.4 ± 0.7 a	8.4 ± 0.3 a
	Control	13.5 ± 10.3	6.1	3.3	12127.9 ± 1168.8 a		16	14	60	4.0	69.9	14.4 ± 0.7 a	8.3 ± 0.2 a
3	*IT8*	8.1 ± 2.2	5.5	4.7	**48989.5 ± 5473.4 a**	257	277	60	58	2.2	17.7	30.7 ± 0.3 a	8.1 ± 0.1 a
	IT11	12.5 ± 5.8	7.9	7.5	41898.0 ± 6359.7 b		248	49	76	2.7	16.9	31.1 ± 1.0 a	7.9 ± 0.3 a
	IT13	15.2 ± 8.5	8.1	10	41874.3 ± 4943.6 b		247	42	77	3.2	17.0	30.6 ± 0.6 a	8.2 ± 0.2 a
4	*IT10*	10.1 ± 1.1	7.2	6.6	36352.3 ± 4971.2 a	392	412	88	43	2.2	8.8	20.9 ± 0.8 a	10.3 ± 0.4 a
	IT17	18.5 ± 3.8	10.1	9.8	**29799.3 ± 4791.4 b**		367	71	49	2.8	8.1	21.0 ± 0.6 a	10.2 ± 0.4 a
	IT25	26.4 ± 3.9	14.7	15.4	**30452.9 ± 4909.0 b**		298	52	53	3.8	10.2	20.1 ± 0.8 a	10.4 ± 0.3 a
	Control	7.9 ± 3.1	5.5	3.8	35780.2 ± 5194.8 a		542	62	~75[b]	3.2	6.6	21.2 ± 0.4 a	10.3 ± 0.7 a

Note: The treatments with the same letter do not differ at the 0.05 significance level.

[a] For clarity, the cumulative marketable yield during the reference period for each site is presented here, but the statistical analyses were conducted on individual harvests throughout the season with a repeated measure mixed model. At all Sites, the interaction term between harvest dates and yield was not significant at the 0.05 significance level.

[b] From personal communication with the producer, this is an approximation since the producer did not keep track of irrigation durations in his irrigation log.

2011. At these sites, the harvest period began on July 10th, but the reference period was shortened because data regarding the applied water volumes were unavailable before August because of a water totalizer malfunction. At Sites 3 and 4, the reference periods were from February 15th, 2012 to June 15th, 2012 and from April 1st, 2013 to October 15th, 2013 respectively.

2.10. Marketable yield, fruit quality and plant growth parameters

At the Group A sites, marketable yields from the total plot area were collected on a bi-weekly basis and four fruit subsamples per plot were used to measure the average fruit size and fruit soluble solid content (Brix index) using a refractometer. At the Group B sites, the same information was gathered from 2 weekly harvested sub-plots per each experimental plot. At the latter sites, 2–5 plants per sub-plot were selected to obtain weekly measurements of crown diameter and canopy coverage area. More information on the configuration of sub-plots is provided in Table 2. The plots at the Group B sites were normally harvested several hours before commercial fruit harvest. On some occasions, coordination problems between the harvest crew and the research team occurred and the fruits were collected without sorting according to the experimental treatments. Hence, the cumulative marketable yield at these sites represented only a fraction of the total yield for the reference period.

2.11. Irrigation parameters

At all sites, a water totalizer was installed on the supply line for the drip laterals associated with each treatment and the applied irrigation water volume was measured weekly during the reference period. WUE was calculated by dividing the marketable yields by the volume of applied water. An irrigation log was used to keep track of the frequency and duration of the irrigation events and the ψ measured at a depth of 15 cm at irrigation initiation was used to calculate the average ITs achieved for the reference period. The reference ET (ET_0) was calculated using climatic data from an on-site or nearby weather station and the FAO modified Pennman–Montheith equation and converted to crop ET (ET_c) by using appropriate crop coefficients according to canopy coverage (Grattan et al., 1998) and/or the number of days after planting (Hanson and Bendixon, 2004).

2.12. Soil physical properties

Conventional loose soil samples and undisturbed core samples were collected at depths of 15 and 30 cm from the growing beds during the first weeks following planting. Conventional samples were used to measure the soil texture using the hydrometer method (Gee and Or, 2002a) [Sites 1 and 2] or the laser light scattering method (Gee and Or, 2002b) [Sites 3 and 4]. In addition, the soil electrical conductivity was determined using the 1:1 saturated soil paste extract method (Rhoades, 1982). Core samples were used to measure the saturated hydraulic conductivity (K_{sat}) using the constant head method (Reynolds et al., 2002). Desorption water retention curves were obtained from core samples with suction tables and pressure plate extractors (Dane and Hopmans, 2002a and Romano et al., 2002) for Sites 1 and 2 and with pressure cells (Dane and Hopmans, 2002b) for Sites 3 and 4. The soil bulk density was determined by oven drying and weighing the core samples. For the Group A sites, samples were collected from 27 locations across a regularly spaced grid to assess the spatial variability of the soil properties by using basic (experimental variogram only) geostatistical analysis (Yates and Warrick, 2002). For the Group B, samples were collected at 2–3 locations per plot for a total of 12 and 15 samples at Sites 3 and 4 respectively.

2.13. Soil salinity measurements

At all sites, weekly measurements of the electrical conductivity and pH of the soil solution were performed on samples extracted using suction lysimeters to insure that they did not exceed the recommended values for strawberry production (Barroso and Alvarez, 1997). For the same purpose, soil samples were collected monthly at a depth of 15 cm in all plots to measure the soil electrical conductivity by using the 1:1 saturated paste extraction method (Rhoades, 1982).

2.14. Statistical analysis

To statistically analyze the yield, fruit size and soluble solid content, an analysis of variance test was performed using a linear mixed-effects repeated measures model with the *nlme* package in R (Pinheiro et al., 2013). The main effect contrasts were determined using the *multcomp* package (Hothorn et al., 2008), and a 0.05 significance level was used in all tests. For the Group B sites, the yield data from the first harvest were discarded from the statistical analyses because not all of the plants began producing fruit simultaneously and because early variability results from uncontrollable differences in plant size and vigor that were present at the moment of planting or occurred due to the adequacy of the planting technique (crown depth, root damage).

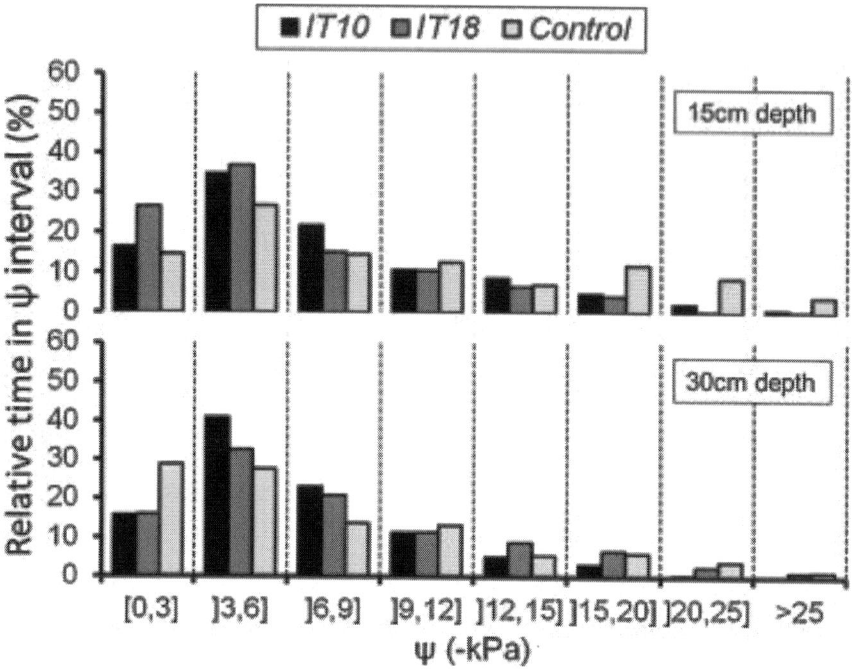

Figure 3. Relative time spent within the different matric potential ranges during the reference period of the experiment at Site 1.

3. RESULTS AND DISCUSSION

3.1. Site 1: Soil matric potential

Abundant and frequent precipitations, soil water retention properties and fertigation events conducted by the producer made the achievement of the intended IT difficult at Site 1. Hence, the applied treatments did not result in major differences in the soil matric potential during the reference period (Fig. 3). For *IT10* and *IT18,* the average achieved ITs were −13.3 and −14.9 kPa rather than the intended −10 and −18 kPa, respectively. The average ψ values at both measurement depths were also similar between the IT treatments. *IT10* was in fact maintained in slightly dryer conditions than *IT18* at a 15 cm depth, but it was maintained slightly wetter at a 30 cm depth (Table 3). For the *Control*treatment, the (unintended) achieved IT was −24.7 kPa and the average ψ at a depth of 15 cm was 2.7–3.9 kPa lower than for *IT10* and *IT18*, respectively (Table 3). The failure to obtain the intended IT for *IT10* can be explained by the water retention curve of this soil (Fig. 4). Similarly to coarse sand, the water retention curve presents a sharp decrease in water content in the wet part of the curve (0 to −5 kPa). However, similar to clay, the water content decreases very slowly when ψ is less than −6 or −8 kPa. Consequently, ψ in this soil decreases rapidly above values of −6 to −8 kPa. This allows little intervention time for initiating irrigation before the threshold is exceeded. For *IT18*, fertigation events conducted by the grower before the IT was reached explain why the achieved IT was higher than intended.

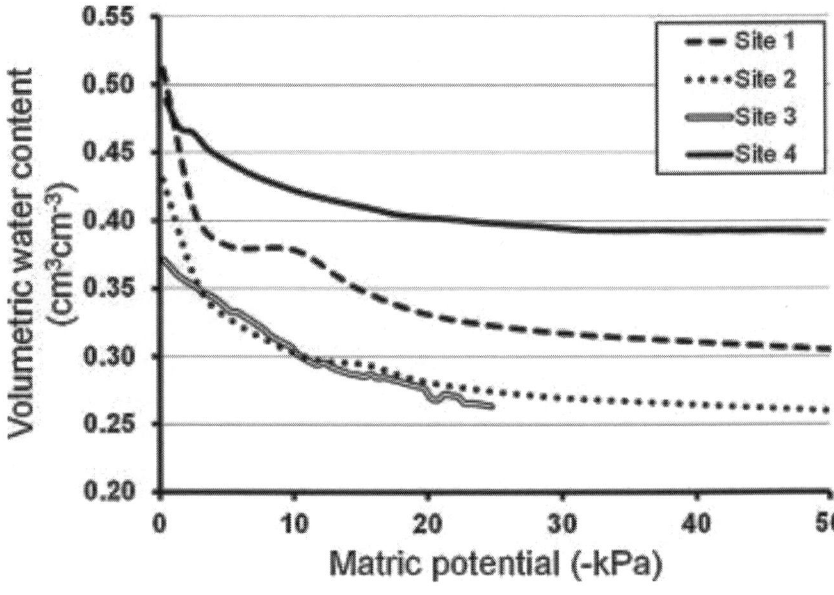

Figure 4. Average water retention curve from the four experimental sites.

3.2. Site 1: Yield and fruit quality

Even if not perfectly achieved, the application of the irrigation treatments resulted in a significant increase in the marketable yield of 6.2% for *IT10* and *IT18* relative to *Control*($p = 0.01354$). However, *IT10* and *IT18* did not significantly differ from one another (Table 3). The statistical analysis also showed that the interaction term between harvest date and yield was not significant ($p = 0.3029$). Hence the yield increase achieved with both IT treatments relative to *Control* remained constant during the reference period. No significant difference in fruit quality or size was observed between any of the treatments (Table 3). Therefore, the yield improvements obtained with the

IT treatments were due to differences in the number of fruits produced by the plants.

The results regarding the influence of the treatments on ψ and on yield can be used to estimate the optimal IT range for this Site. Fig. 3 shows differences between *IT10* and*IT18* with regards to the relative time spent with ψ at a 15 cm depth in the [0,−3], and [−6, −9] kPa ranges. It is particularly clear that *IT18* was maintained above −3 kPa for longer durations than *IT10*. However, the same yields were achieved with both IT treatments. This suggests that maintaining the bed under wet conditions (>−3 kPa) do not improve fruit production. Fig. 3 also shows that ψ in all treatments was between −9 and −15 kPa for similar durations; and that the relative time in the [−15, −20] kPa range was 163% more important for *Control* than for the IT treatments. Hence, the hydric stress that reduced yield for *Control* potentially occurred when ψ was lower than −15 kPa. If only yields were to be considered, −15 kPa would thus be an adequate IT at this site.

3.3. Site 1: Water use efficiency

Regarding water application, the *IT18* and *Control* treatments required similar volumes of water, whereas the *IT10* treatment resulted in a 136% increase in water application relative to the other two treatments (Table 3). The *IT10, IT18* and *Control* plots were irrigated 19, 9 and 6 times during the reference period, and the average irrigation durations were 40, 48 and 60 min, respectively. Hence, even when not perfectly achieved, the intended ITs and the irrigation duration adjustment procedure had notable impacts on the applied water volume and on the irrigation frequency and duration. The WUE was better in the *IT18* and *Control* treatments despite significant total yield decrease associated with the latest. By definition, WUE is related to yield and applied water volumes only. In this case, it does not account for any increased application efficiency in *IT18* relative to the *Control*. This increase in efficiency occurred because monitoring the matric potential under the root zone most likely prevented excessive leaching. This is supported by the fact that ψ at a depth of 30 cm was more frequently within the [0, −3] kPa range (above bed capacity) in *Control* than in *IT18* (Fig. 3). This indicates that gravitational leaching was most likely more frequent in *Control* than in *IT18.IT18* was only irrigated on 3 more occasions than *Control*, and the same total amount of water was applied in both treatments, which demonstrates the critical issue of irrigation timing. The additional water applied to *IT10* relative to *IT18* did not affect yield but substantially reduced the WUE. Despite being classified as a silty clay loam, the soil from Site 1 had a an important proportion of rock fragments and a low bulk density. In agreement with the previously discussed water retention properties this suggests that macropores represent an important part of this soil's poral network, resulting in a drainage capacity much better than expected for a silty clay loam. In these conditions, it is likely that the excess water did not generate anoxic conditions that could be detrimental to the plants. In summary, the results from this site suggest that an IT of approximately −15 kPa could optimize strawberry yield and WUE (Fig. 5). This result is consistent with the optimal values reported for a sandy soil with a high organic matter content (Hoppula and Salo, 2007) and with the values reported by Bergeron (2010) for a nearby field.

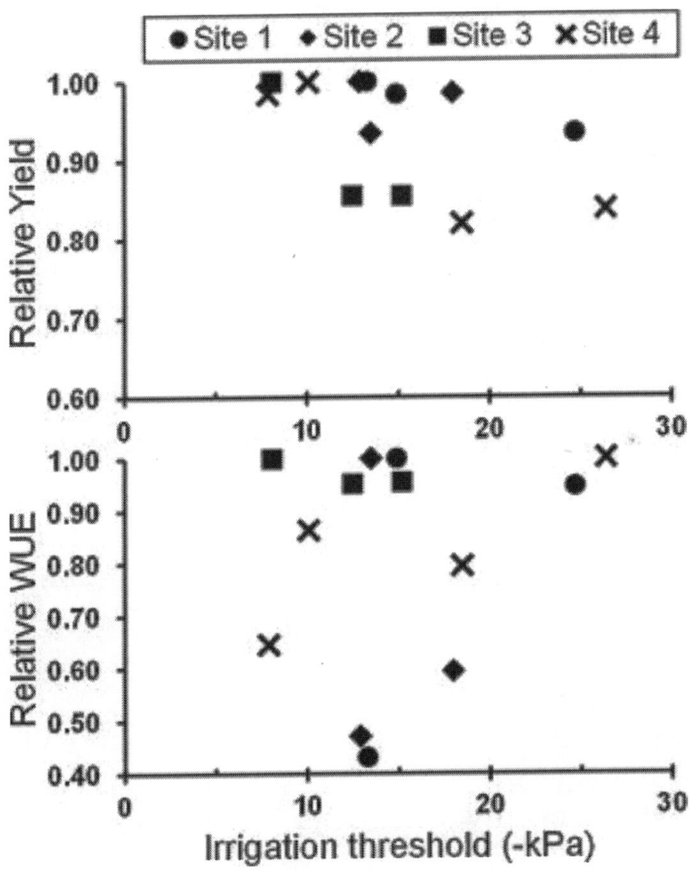

Figure 5. Effect of the irrigation threshold on the within site relative yield (upper plot) and relative water use efficiency (lower plot) for all experimental sites.

3.4. Site 2: Soil matric potential, yield and fruit quality

At Site 2, the intended and achieved irrigation thresholds agreed well. Nonetheless, the average ψ values at both measurement depths were very similar for all treatments (Table 3), and the differences between the treatments with respect to the relative time spent in different ψ ranges were minimal (Fig. 6). Statistical analysis of the yield data indicated no significant difference between the treatments, but orthogonalcontrasts revealed that the 6.6% yield increased achieved with *IT10* relative to *IT18* was nearly significant ($p = 0.0561$). Fig. 6 shows that *IT10* spent half as much time as *IT18* in the [−9, −12] kPa range. In addition, ψ was maintained lower than −15 kPa over similar durations in both treatments. This result suggests that a yield-affecting hydric stress could occur between -9 and −12 kPa. However, since the yields were not significantly different, additional experiments are needed to confirm this finding. Again, no differences in fruit sugar content or size were observed among any of the treatments (Table 3).

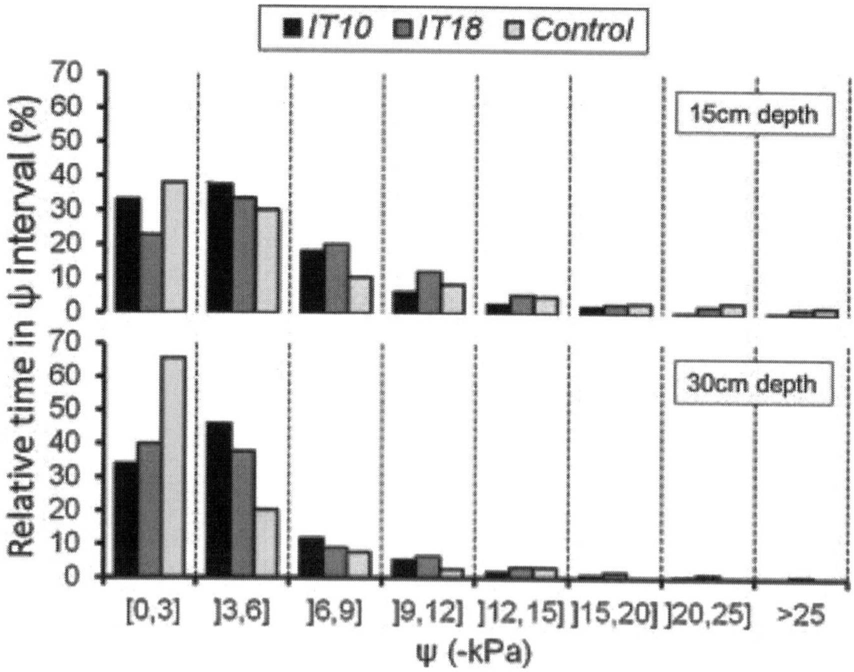

Figure 6. Relative time spent within different matric potential ranges during the reference period of the experiment at Site 2.

Observations subsequently made on Sites 3 and 4 could explain the absence of significant differences at Site 2. At Sites 3 and 4, it was observed that the yields rapidly decreased when a certain IT was exceeded (Fig. 5). For lower ITs, the effects of a decrease in IT on yield were less evident. It is possible that at Site 2, all treatments exceeded the optimal IT and were within or near the «plateau» part of the yield response curve to IT. This hypothesis is supported by unsaturated hydraulic conductivity measurements made at a nearby site during the following season, which suggested that the water transfer rate in the soil matrix might become limiting between −5 and −8 kPa on days with high evapotranspiration demands. However, additional experiments with higher irrigation thresholds are needed to confirm this finding.

3.5. Site 2: Water use efficiency

Relative to *Control, IT10* and *IT18* led to increases in applied water volume of 128% and 78% respectively. This greatly reduced their WUE (Table 3). Tensiometer-based irrigation has also been reported to increase the volume of applied water (from 47 to 267%) relative to the climatic water balance management in a similar experiment when using an IT of −20 kPa (Krüger et al., 1999). The *IT10, IT18* and *Control* plots were irrigated 20, 15 and 14 times during the reference period, and the average irrigation durations were 40, 42 and 60 min, respectively. The reduction of irrigation durations for both IT treatments relative to *Control* most likely reduced the fraction of the applied water that was leached under the root zone. Three phenomena could explain the poor WUE of tension-based irrigation in this case. First, the water application method was likely inadequate for the soil type. Although the soil at Site 2 was classified as a silty clay loam, it presented saturated hydraulic conductivity values (Table 4) and water retention properties (Fig. 4) that were similar to those of coarse sand. Consequently, the downward movement of water was very fast, with minimal lateral water movement. Thus, most of the applied water was rapidly leached to the soil below the root zone and became unavailable to plants, which would explain the limited effects of irrigation.

Table 4. Basic physical properties of the soils from the experimental sites, where ρ_{bulk} is the soil bulk density, andK_{sat} is the hydraulic conductivity.

Site		Sand content (%)	Silt content (%)	Clay content (%)	Particle > 2 mm content (%)	Organic matter content (%)	Soil depth (cm)	ρ_{bulk} ($g\,cm^{-3}$)	K_{sat} 15 cm depth ($cm\,h^{-1}$)	K_{sat} 30 cm depth ($cm\,h^{-1}$)
1	Avg	18.1	48.4	33.5	32.8	5.7	20–30	1.04	158.8	67.0
	CV (%)	*41.3*	*12.0*	*15.7*	*37.9*	*16.8*		*9.8*	*69.7*	*114.2*
2	Avg	28.4	44.8	26.8	NA	2.8	30–40	1.23	115.1	55.8
	CV (%)	*23.2*	*9.9*	*12.7*	*NA*	*24.2*		*8.4*	*61.1*	*148.8*
3	Avg	60.0	27.9	12.1	<1	1.2	> 100	1.53	1.3	13.1
	CV (%)	*3.0*	*7.0*	*5.1*	*NA*	*9.6*		*4.0*	*79.7*	*140.9*
4	Avg	37.6	32.5	29.9	<1	2.4	> 100	1.32	44.7	NA
	CV (%)	*7.4*	*4.6*	*6.3*	*NA*	*11.8*		*11.8*	*188.2*	*NA*

Second, it is possible that the abundant precipitation during the reference period, combined with the presence of a water-movement-restricting layer under the monitored depth of 30 cm, allowed the soil to maintain a sufficient water reserve for the plants. During the reference period, the strawberry plants most likely had a root system that reached 30 cm and could benefit from such a reserve. This hypothesis is supported by the observation that the matric potential at a depth of 30 cm was maintained near bed capacity during most of August and September. Because of the plastic mulches over the beds, rainwater does not significantly affect the matric potential at 15 cm and the poor capillary rise capacity of the soil from Site 2 does not favor recharge from the bottom soil layers. It is possible that although sufficient water for root uptake was available at greater depths, the matric potential at 15 cm continued to decrease, which triggered irrigation.

Finally, the short scale spatial variability of soil properties could explain the absence of a yield response to irrigation regimes. Even if the variographic analysis showed no spatial structure at the investigated measurement scale at the site (data not shown), the coefficients of variation of the soil properties (Table 4), especially the sand and organic matter contents, were important. This most likely influenced the response of the matric potential to irrigation and root uptake. For the IT treatments, irrigation was triggered based on the average ψ value obtained from two monitoring stations. However, at the time of irrigation initiation, substantial discrepancies were often observed between ψ measurements from the two stations for both *IT10* (Fig. 7) and *IT18* (not shown). None of the soil properties measured near the monitoring stations justified such a difference in hydraulic behavior. However, it is possible that coarse particles (> 2 mm), such as schist fragments, were located near some of the tensiometers. In such cases, rapid drainage would be favored as the water retention capacity decreased, which would require more frequent irrigation. Variographic analysis of the proportion of coarse particles at a nearby site (Letourneau 2015, unpublished) indicated a significant nugget effect and no spatial structure at any scale. Hence, random effects of coarse particles potentially altered the effects of the applied treatments.

3.6. Site 3: Spatial variability of the matric potential

Table 5 compares the mean *ψ* values from the manual tensiometers that were spread across the experimental site with the values obtained from the monitoring station. It also presents the results from the Student t-tests that were performed to verify if they were significantly different. Out of 18 data sets, the measurements with both methods were significantly different on only two occasions. In more than 50% of the cases, the differences between the matric potentials were less than 1 kPa. In addition, these comparisons were performed by the end of the season when the effects of variability were likely highest because plant development and root uptake would have been affected by the differences in soil water status for 7 months. At the scale of this experiment (i.e. for an area of 0.25 ha per treatment), tension measurements from a single monitoring station in one replicate were representative of observations in all 3 replicates. In this case, the soil from this site was very uniform, as shown by the low coefficients of variation of most of the soil properties (Table 4). More experiments would be required to determine the number of monitoring stations required for larger plots or for less uniform soils.

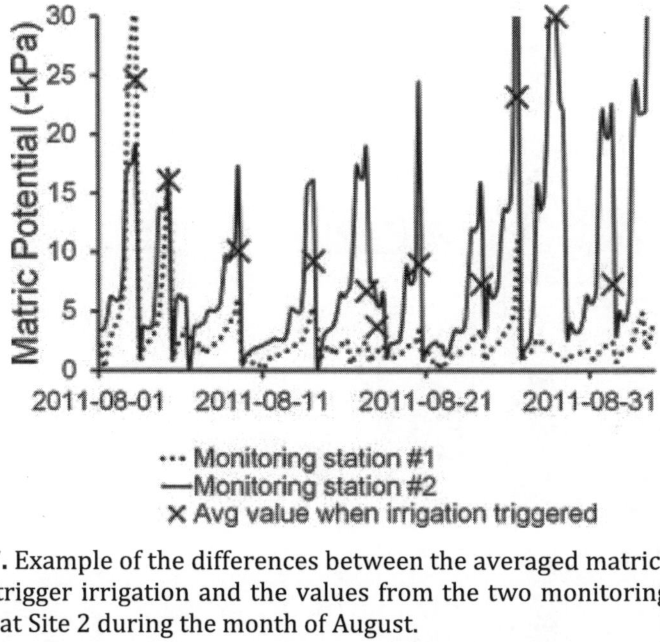

Figure 7. Example of the differences between the averaged matric potential used to trigger irrigation and the values from the two monitoring stations for IT10 at Site 2 during the month of August.

Table 5. Results from the Student t-test procedure used to compare the matric potential observations from the monitoring station and the mean value from the manual tensiometers, where ψ is the matric potential at 15 cm.

Measurement run	*IT8*				*IT11*				*IT13*			
	ψ Monitoring station (−kPa)	Avg ψ manual probes (−kPa)	Degrees of free-dom	*p* Value of *t*-test	ψ Mon-itoring station (−kPa)	Avg ψ manual probes (−kPa)	Degrees of free-dom	*p* Value of *t*-test	ψ Mon-itoring station (−kPa)	Avg ψ manual probes (−kPa)	Degrees of free-dom	*p* Value of *t*-test
1	5.2	4.3 ± 2.6	5	0.4429	7.8	6.2 ± 0.8	3	**0.0289***	6.6	6.2 ± 4.0	3	0.8453
2	7.6	7.1 ± 1.8	4	0.5921	12.6	9.6 ± 4.5	3	0.2690	9.7	10.7 ± 4.9	3	0.7222
3	5.9	5.4 ± 3.0	5	0.7172	6.9	6.3 ± 1.4	3	0.4632	4.6	7.7 ± 3.5	3	0.1790
4	7.0	2.9 ± 1.3	4	**0.0021***	8.8	6.9 ± 1.3	2	0.1169	7.0	6.4 ± 3.5	4	0.7013
5	4.5	3.6 ± 1.6	5	0.2099	7.9	5.2 ± 2.4	3	0.1041	9.3	11.9 ± 4.5	3	0.3368
6	5.0	4.6 ± 1.0	4	0.3936	5.7	4.0 ± 1.8	4	0.1013	10.7	10.7 ± 7.4	4	0.995

3.7. Site 3: Soil matric potential

Except for the end of the production season, when ETo reached values of up to 6 mmd^{-1}and the ITs were occasionally exceeded for all treatments, the agreement between the intended and achieved ITs was good (Table 3). As shown in Fig. 8, *IT8* was maintained at a ψ above −9 kPa at both measurement depths during 95% of the reference period. Respectively, *IT11* and *IT13* remained above −9 kPa for 72 and 66% of the reference period. Both *IT11* and *IT13* spent equivalent amounts of time between −9 and −12 kPa at both depths. As shown by the average *ψ* values at both depths (Table 3) and in Fig. 8, the most noticeable difference between *IT11* and *IT13* was that the latter was maintained in drier conditions at a depth of 30 cm. A greater difference in IT between *IT11* and *IT13*would likely be required to induce notable differences regarding the amount of time spent in different *ψ* ranges. Overall, it was not very practical to achieve a separation of only 2 kPa without using an automated irrigation system.

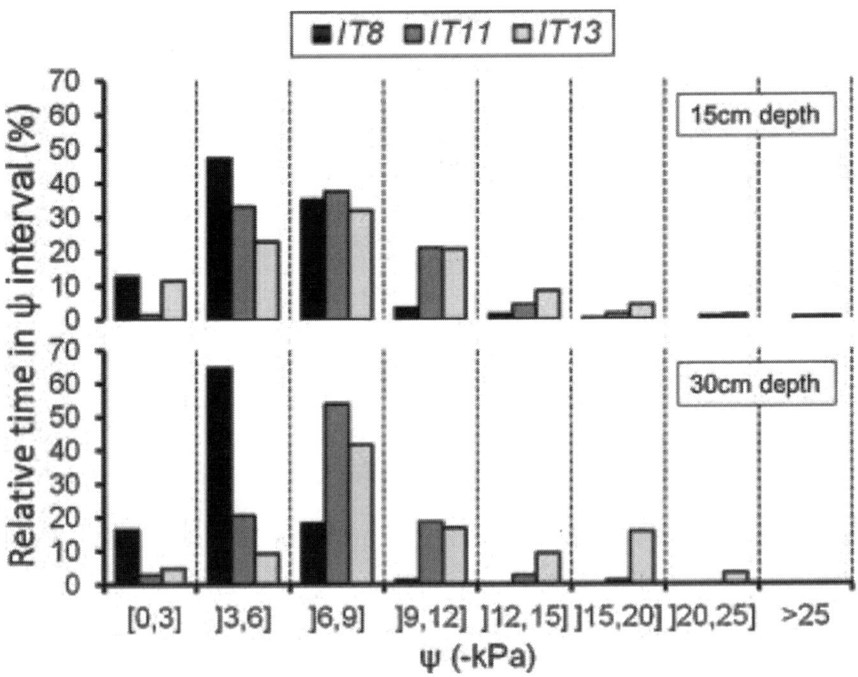

Figure 8. Relative time spent within different matric potential ranges during the reference period of the experiment at Site 3.

3.8. Site 3: Yield and fruit quality

With respect to the marketable yield, orthogonal contrasts indicated that *IT*8 resulted in a significant ($p = 0.0434$) 17% marketable yield increase relative to *IT*11 and *IT*13 (Table 3). This difference in yield roughly corresponded to the amount of time spent within the [−9, −12] kPa range at both depths for *IT11* and *IT13* (19%), which suggested that irreversible yield-affecting stress principally occurs within this range. No differences between the treatments regarding the fruit sugar content and size were observed. Thus, yield differences were attributable to a difference in the fruit number. As for Site 1, the interaction term between harvest date and yield was not significant ($p = 0.6473$). Hence the difference in yield between *IT8* and the other treatments remained constant during the reference period. These results are consistent with previous studies. Optimal yield was obtained with an IT of −10 kPa for production under high-tunnel in a similar soil (Guimerà et al., 1995). Regular EC measurements from soil saturated paste extracts and suction lysimeter samples showed that soil salinity remained at similar levels for all treatments and did not reach values that could have affected plant growth. For all treatments, SAR was less than 2.0, which is much lower than the critical value (6.0) for berry production. This result differed according to the position within the growing bed, which was expected in a drip irrigation context (Hanson and Bendixon, 2004) but not between treatments. Knowing that the same fertilization method was applied to all treatments and that the salinity most likely did not affect the yields, we can confidently assert that the yield differences between the treatments mostly resulted from water availability.

3.9. Site 3: Water use efficiency

The application of water to *IT*11 and *IT*13 agreed well with the calculated ETc for the reference period (257 mm), while *IT*8 required the application of 7.4% more water than ETc (Table 3). The WUE of *IT8* was better than the WUEs in the other treatments because the yield increase was greater. Regarding the irrigation frequency and duration, irrigation was required on average every 2.2, 2.7 and 3.2 days for durations of 58, 76 and 77 min for *IT*8, *IT*11 and *IT*13, respectively. Such a difference in yield with an irrigation frequency difference of less than one day shows that daily measurements are not sufficient for optimizing strawberry irrigation management, regardless of the criterion used (climatic-, soil- or plant-based measurements). This finding is consistent with reports from Jones (2004) that real-time or hourly measurements are required for optimal irrigation scheduling. Fig. 8 shows that *IT8* was maintained between 0 and

−3 kPa at a depth of 30 cm for 16% of the reference period, which is 4 times more than *IT11* and IT13. In this case, more leaching likely occurred under the root zone. This finding suggests that the method used to adjust the duration of irrigation events could require some adjustments for soils with such low hydraulic conductivities (Table 4). This soil's poor water transfer capacity mainly occurred because 80% of its sand particles were smaller than 0.25 mm (classified as fine to very fine sand), which increased the proportion of flow restricting micropores. Due to slow water infiltration, it is difficult to adjust irrigation durations to obtain target matric potentials at both measurement depths simultaneously. In Summary, the experiment at Site 4 showed that triggering irrigation at −8 kPa provided optimal yields and WUE. They also suggest that high-frequency and short-duration irrigation could minimize leaching under the root zone in these conditions, thus further improve WUE.

3.10. Site 4: Special considerations

Except for the K_{sat} measurements in which variability is frequently elevated because of sampling and measurement errors (Reynolds et al., 2002), the soil properties from Site 4 had low variation coefficients. Thus, the soil was considered as uniform. Because the experimental area was nearly 3 times smaller than that at Site 4, it was assumed that measurements from a single monitoring station per treatment would represent the ψvalues across the entire area.

Similar to many other fields in the area, the experimental site was affected by cyclamen mites that affected the general plant vigor by feeding on the plant material and altered flower pollination and reduced the fruit size. One replicate of the *IT17* treatment was particularly affected and removed from the statistical analysis. However, all of the other experimental plots were equally affected. On average, approximatively 10,000 kg/ha of berries were misshaped or smaller than industry standards and classified as unmarketable, with no significant differences between the treatments. In addition, canopy coverage was reduced relatively to normal values in the area, which reduced the ETc. No significant difference in canopy coverage was observed between any of the treatments. When calculated with Kc values derived from canopy coverage as described by Grattan et al. (1998), the ETc value for the reference period was 392 mm. However, when calculated using historical monthly adjusted Kc values, the ETc value was 467 mm.

3.11. Site 4: Soil matric potential, yield and fruit quality

At Site 4, the achieved and intended thresholds were in good agreement; and average ψvalues at both depths confirmed that IT treatment application was successful (Table 3). This was facilitated by the low rainfall (24 mm) and regular evapotranspirative demand during the reference period. As intended, decreasing the IT from −10 to −25 kPa resulted in notable differences in the time spent within the different ψ ranges (Fig. 9). Managed by the producer without a pre-determined IT, the *Control* treatment was maintained under the wettest conditions. Consequently, orthogonal contrasts showed that *IT10* and *Control*led to a significant ($p = 0.0111$) 19.7% yield increase relative to *IT17* and *IT25* (Table 3). No significant differences were observed regarding fruit sugar content and size. Again, the interaction term between harvest date and yield was not significant ($p = 0.7180$). The effects of treatments on yield (Table 3) combined with their impacts on the soil water status (Fig. 9) can be used to deduce the optimal ψ range for fruit production. First, under wet conditions, large differences between *IT10* and *Control* regarding the amount of time spent in the [0,−3], [−3,−6] and [−6,−9] kPa ranges did not affect yield. The yields in*Control* were not increased relative to that of *IT10* by spending a greater proportion of the reference period at a ψ above −6 kPa. Second, ψ at a depth of 30 cm in *Contro*l was maintained above the bed capacity (−3 kPa) for as much as 33% of the reference period, which implies that an important fraction of the applied water was leached under the root zone. This excessive water application did not have any adverse effects on fruit production and is consistent with physiologically oriented observations indicating that strawberry plants can maintain their photosynthetic activity while flooded by partially closing their stomata (Blanke and Cooke, 2004). Third, all IT treatments spent similar amounts of time within the [−9,−12] kPa range without any negative effects on yield for*IT10*. Finally, considering both measurement depths, the relative time below −15 kPa was nearly 4 times greater in *IT25* than in *IT17* without any effect on yield. Together, these observations suggest that [−6,−12] kPa is the optimal range in this soil for strawberry yields, and that most of the yield-affecting stress occurs in the range of [−12,−15] kPa. In this case, the time spent in the [−15, −25] kPa range did not affect the yield. However, more experiments would be required to determine if drier conditions would induce a gradual or stepwise yield decrease.

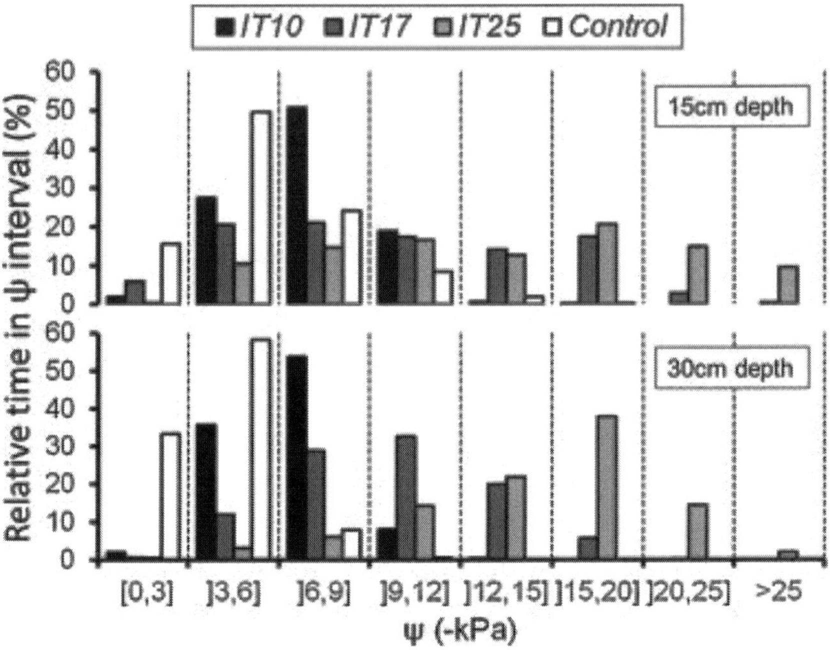

Figure 9. Relative time spent within different matric potential ranges during the reference period of the experiment at Site 4.

3.12. Site 4: Water use efficiency

With regard to water use, *IT10*, *IT17*, *IT25* and *Control* plots were irrigated 88, 71, 52 and 62 times during the reference period, and the average irrigation durations were 43, 49, 53 and approximatively 75 min, respectively (Table 3). The applied irrigation water depths were 105, 94 and 76% of ETc for *IT10*, *IT17* and *IT25*, respectively. For *Control*, 138% of ETc was applied; however, irrigation was used in the control to compensate for the historic ETc, which, was expected to be higher, as discussed above (Table 3). In fact, by applying 532 mm of irrigation water during the reference period, the grower applied 116% of the expected value (467 mm) based on historic monthly means. Considering the water application efficiency and possible salt leaching needs, this amount of water would most likely have been appropriate for the same reference period during an average growing season (i.e. for healthy strawberry plants). However, in this case, the additional water applied to *Control* resulted in a decreased WUE relative to *IT10*. The WUE decreased from *IT10* to *IT17*, mainly because of a yield reduction, and increased between *IT17* and*IT25* mainly due to a reduction in the amount of water applied. Even in a water scarcity situation, it is not likely that growers would sacrifice yield to improve WUE, which was the case for *IT25*. From a practical perspective, it would be interesting to conduct further experiments to identify a point between *IT10* and *IT17* that would allow for a compromise between yield and WUE.

These results show the adaptability of ψ-based irrigation management. Without any special considerations during treatment application, the water applied to *IT10* and *IT17*was very similar to the calculated ETc (Table 3). By contrast, adjusting the amount of water applied to *Control* in order to compensate the actual water requirement of insect affected plants (which was not an objective in this study), would have required monthly or bi-monthly canopy coverage measurements. These measurements are common for research purposes but are rarely performed in a production context. In this experiment, the ETc was lower than anticipated due to mites. However, these observations suggest that matric potential-based irrigation could automatically (i.e. without any additional measurements) account for most uncontrolled (pest, disease, weather) or planned (varieties, plant density) factors that could affect plants because it is directly related to root uptake. In addition, on a volumetric basic, the water applied to *IT10* and *IT17* was very similar to the ETc estimations but resulted in major differences in yield. Again, this stresses the importance of irrigation timing. In this case, proper timing allowed *IT10* to remain above the presumed critical ψ value of −12 kPa, leading to better yield relative to IT17. In summary, ETc calculations have been shown here and in many other experiments to serve as a very interesting tool for determining the amounts of water required by plants. However, to optimize irrigation management, ETc should be used in combination with another criterion available over a short time interval. At this site, real-time monitoring of the matric potential was shown to be an efficient criterion.

4. CONCLUSION

Experiments conducted at the Group A sites indicated that matric potential-based irrigation management could positively affect the marketable yield and WUE, even for short production periods with abundant precipitation (Fig. 5). These experiments also allowed to identify potential limitations of that type of management. These limitations were (1) the difficulty of achieving intended thresholds in production conditions, (2) the effect of the unstructured spatial variability of soil properties and (3) management difficulties associated with inadequate wetting patterns from the SDI irrigation system. Additional experiments designed to understand and differentiate the effects of these limitations would be required to optimize matric potential-based management under these conditions. The results from Sites 1 and 2 may be useful for improving irrigation management practices on farms located in the same areas as the experimental sites. Unfortunately, because of the specific properties (schist fragments) of the soils at these sites, it is not appropriate to use information regarding the yield responses to IT that were obtained from these experiments to infer optimal thresholds for soils with similar textures because conventional textural classification only includes particles smaller than 2 mm. Knowledge of more complex properties, such as water retention and hydraulic conductivity, or conducting similar experiments on more typical soils from different soil types would be required to apply these results to other sites.

At the Group B sites, which have uniform soils and nearly no precipitation, the effects of IT management on yield and WUE were much more pronounced (Fig. 5). Initiating irrigation at a proper matric potential allowed to improve the yield and WUE, respectively, by as much as 20 and 33% relative to conventional irrigation practices in this area. In general, the yields from the Group B sites were not negatively affected by high matric potentials (wet conditions), for which aeration could be limiting. However, the yield decreased sharply for matric potentials of less than −8 to −12 kPa, depending on the site. In dryer conditions, the yield response to IT appeared to reach a plateau for IT values between −12 and −25 kPa depending on the site. This suggests that the soil matric potential range for which a compromise between optimal yield and WUE is possible would be between −10 and −15 kPa, depending on the site. At Site 4, leaching under the root zone was also considerably reduced relatively to conventional practices by monitoring the matric potential at a depth of 30 cm following irrigation. Of course, additional experiments could result in further optimization of irrigation practices at all of these sites. It would be interesting to verify if high-frequency and short-duration irrigations could reduce leaching while maintaining high level yields. The soil from Site 3 is a typical fine textured sandy loam, and the soil from Site 4 is a typical smectitic clay loam. Both soils are well represented in their respective areas. Thus, the results could be used as a guideline to improve irrigation practices in similar soils. The results from sites with contrasting properties, varieties and cultural practices were similar under temperate (Group A) and warm (Group B) climatic conditions. It is thus likely that an IT of −10 kPa would provide good results in most type of soils and under most climatic conditions. It would at least represent a good starting point for further optimization of irrigation management of field grown strawberries.

ACKNOWLEDGMENTS

The authors would like to express their gratitude to Ferme Onésime Pouliot for providing two experimental sites and for their technical and financial contributions to the research project. This project was funded by the Natural Sciences and Engineering Research Council of Canada (NSERC), Hortau, and Ferme Onésime Pouliot.

REFERENCES

1. Barroso, M.C.M., Alvarez, C.E., 1997. Toxicity symptoms and tolerance of strawberry to salinity in the irrigation water. Sci. Hortic. (Amsterdam) 71, 177–188,
2. Bergeron, D., 2010. Régie de l'irrigation goutte à goutte dans la production de fraises à jour neutres au Québec. Fac. des Sci. l'Agriculture l'Alimentation, département des Sols Génie Agroaliment. Université Laval, Québec.
3. Blanke, M.M., Cooke, D.T., 2004. Effects of flooding and drought on stomatal activity, transpiration, photosynthesis, water potential and water channel activity in strawberry stolons and leaves. Plant Growth Regul. 42, 153–160,
4. Dane, J.H., Hopmans, J.W., 2002a. 3.3.2.4 Pressure Plate Extractor. In: Dane, J.H., Topp, G.C. (Eds.), Methods of Soil Analysis. Part 4-Physical Methods. Soil Science Society of America, Inc., Madison, WI, USA, pp. 688–690.
5. Dane, J.H., Hopmans, J.W., 2002b. 3.3.2.3 Pressure Cell. In: Dane, J.H., Topp, G.C. (Eds.), Method of Soil Analysis, Part 4-Physical Methods. Soil Science Society of America, Inc., Madison, WI, USA, pp. 684–688.
6. Gee, G.W., Or, D., 2002. 2.4.3.5Hydrometer Method, in: Methods of Soil Analysis, Part 4-Physical Methods. Madison, WI, USA, pp. 278–283.
7. Gee, G.W., Or, D., 2002b. 2.4.3.6. c Laser Light Scattering (Diffraction). In: Dane, J.H., Topp Clark, G. (Eds.), Methods of Soil Analysis. Part 4-Physical Methods. Soil Science Society of America, Inc., Madison, WI, USA, pp. 286–288.
8. Giné Bordonaba, J., Terry, L.A., 2010. Manipulating the taste-related composition of strawberry fruits (Fragaria×ananassa) from

different cultivars using deficit irrigation. Food Chem. 122, 1020–1026,

9. Grattan, S.R., Bowers, W., Dong, A., Snyder, R.L., Carrol, J.J., William, G., 1998. New crop coefficients estimate water use of vegetables, row crops. Calif. Agric. 52 (1), 16–21.

10. Guimerà, J., Marfà, O., Candela, L., Serrano, L., 1995. Nitrate leaching and

11. strawberry production under drip irrigation management. Agric. Ecosyst.

12. Environ. 56, 121–135,

13. Hanson, B.R., Bendixon, W., 2004. Drip irrigation evaluated in Santa Maria Valley strawberries. Calif. Agric. 58, 48–53, 10.3733.

14. Hendrickx, J.M.H., Nieber, J.L., Siccama, P.D., 1994. Effect of tensiometer cup size on field soil water tension variability. Soil Sci. Soc. Am. J. 58, 309–315,

15. Hendrickx, J.M.H., Wierenga, P.J., 1990. Variability of soil water tension in a trickle irrigated Chile pepper field. Irrig. Sci. 11, 23–30,

16. Hoppula, K., Salo, T., 2007. Tensiometer-based irrigation scheduling in perennial strawberry cultivation. Irrig. Sci. 25, 401–409,

17. Jones, H.G., 2004. Irrigation scheduling: advantages and pitfalls of plant-based methods. J. Exp. Bot. 55, 2427–2436,

18. Klamkowski, K., Treder, W. 2006. Morphological and Physiological Responses of Strawberry Plants to Water Stress, 2006.

19. Klamkowski, K., Treder, W., 2008. Response to drought stress of three strawberry cultivars grown under greenhouse conditions. J. Fruit Ornam. Plant Res. 16, 179–188.

20. Krüger, E., Schmidt, G., Brückner, U., 1999. Scheduling strawberry irrigation based upon tensiometer measurement and a climatic water balance model. Sci. Hortic. (Amsterdam) 81, 409–424,

21. Pelletier, V., Gallichand, J., Caron, J., 2013. Effect of soil water potential threshold for irrigation on cranberry yield and water productivity. Trans. Am. Soc. Agric. Biol. Eng. 56, 1325–1332,

22. Pe˜nuelas, J., Savé, R., Marfà, O., Serrano, L., 1992. Remotely measured canopy temperature of greenhouse strawberries as indicator of water status and yield under mild and very mild water stress conditions. Agric. For. Meteorol. 58, 63–77

23. Périard, Y., Caron, J., Jutras, S., Lafond, J.A., Houlliot, A., 2012. Irrigation management of romaine lettuce in histosols at two spatial scales: water, energy, leaching and yield impacts. WIT Trans. Ecol. Environ. 168, 171–188,

24. Pinheiro, J., Bates, D., DebRoy, S., Sarkar, D., Team, T.R.C., 2013. nlme: Linear and Nonlinear Mixed Effects Models. R Packag. version.

25. Rekika, D., Caron, J., Rancourt, G.T., Lafond, J.A., Gumiere, S.J., Jenni, S., Gosselin, A., 2014. Optimal Irrigation for Onion and Celery Production and Spinach Seed Germination in Histosols 981–994.

26. Reynolds, W.D., Elrick, D.E., Young, E.G., 2002. 3.4.2.2Constant Head Soil Core (tank) Method. In: Dane, J.H., Topp, C.G. (Eds.), Methods of Soil Analysis. Part 4-Physical Methods. Soil Science Society of America, Inc., Madison, WI, USA, pp. 804–809.

27. Rhoades, J.D., 1982. Soluble Salts, In: Methods of Soil Analysis. Part-2 Chemical and Microbiological Properties. Madison, WI, USA, pp. 167–179.

28. Romano, N., Hopmans, J.W., Dane, J.H., 2002. 3.3.2.6 Suction Table. In: Dane, J.H., Topp, G.C. (Eds.), Methods of Soil Analysis, Part 4-Physical Methods. Soil Science Society of America, Inc., Madison, WI, USA, pp. 692–698.

29. Savé, R., Penuelas, J., Marfa, O., Serrano, L., 1993. Changes in leaf osmotic and elastic properties and canopy structure of strawberry under mild water stress. Hortscience 28, 925–927.

30. Serrano, L., Carbonell, X., Savé, R., Marfà, O., Pe˜nuelas, J., 1992. Effects of irrigation regimes on the yield and water use of strawberry. Irrig. Sci. 13, 45–48,

31. Shock, C.C., Wang, F.X., 2011. Soil water tension, a powerful measurement for productivity and stewardship. HortScience 46, 178–185.

32. Simonne, E., Studstill, D., Hochmuth, R.C., Olczyk, T., Dukes, M., Munoz-Carpena, R., Yuncong, C.L., 2012. Drip Irrigation: The BMP Era - An Integrated Approach to Water and Fertilizer Management for Vegetables Grown with Plasticulture. Univ. Florida IFAS Ext.

33. Terry, L.A., Chope, G.A., Bordonaba, J.G., 2007. Effect of water deficit irrigation and inoculation with Botrytis cinerea on strawberry (Fragaria×ananassa) fruit quality. J. Agric. Food Chem. 55, 10812–10819,

34. Topp, C.G., Ferré, T.P.A., 2002. 3.1 Water content. In: Dane, J.H., Topp, C.G. (Eds.), Methods of Soil Analysis, Part 4-Physical Methods. Soil Science Society of America, Inc., Madison, WI, USA, pp. 417–546.

35. Yates, S.R., Warrick, A.W., 2002. 1.5 Geostatistics, In: Dane, J.H., G, T.C. (Eds.), Methods of Soil Analysis, Part 4-Physical Methods. Soil Science Society of America, Inc., Madison, WI, USA, pp. 81–116.

36. Young, M.H., Sisson, J.B., 2002. 3.2 Water potential. In: Topp, C.G., Dane, J.H. (Eds.), Method of Soil Analysis, Part 4-Physical Methods. Soil Science Society of America, Inc., Madison, WI, USA, pp. 547–670.

37. Yuan, B., 2004. Effect of drip irrigation on strawberry growth and yield inside a plastic greenhouse. Biosyst. Eng. 87, 237–245,

6

CHAPTER

Dynamics and Activity of Sulfate-Reducing Bacterial Populations in Paddy Soil under Subsurface Drainage: Case Study of Kamboinse in Burkina Faso

Cécile Harmonie Otoidobiga[1*], Amadou Keita[2], Hamma Yacouba[2], Alfred S. Traore[1], Dayéri Dianou[3]

[1]*Research Center for Biological, Alimentary and Nutritional Sciences, Research and Training Unit, Life and Earth Sciences, University of Ouagadougou, Ouagadougou, Burkina Faso*

[2]*International Institute for Water and Environmental Engineering, Ouagadougou, Burkina Faso*

[3] *National Center for Sciences and Technology Research, Ouagadougou, Burkina Faso*

ABSTRACT

Sulfide toxicity is a common disease generally associated with iron toxicity which occurs in rice fields when the Sulfate-Reducing Bacteria (SRB) produce sulfides ions in anaerobic conditions. The high quantity of sulfides ions in the soil solution upsets the mineral element balance in the rice, affects its growth and causes crop yield losses. In Burkina Faso, many rice field soils are abandoned due to sulfides toxicity. The present study was developed to evaluate the impact of subsurface drainage on SRB dynamics and activity during rice cultivation and the incidence on rice production. Twelve concrete microplots with a clay-loam soil and a rice variety susceptible to sulfides toxicity (FKR 19) were used for the experiment. Soil in microplots was drained for 7 days (P1), 14 days (P2), and 21 days (P3), respectively. Control (T) microplots without drainage were prepared similarly. The evolution of SRB populations and the content of sulfides ions in the paddy soil and in soil near rice roots were monitored throughout the cultural cycle using MPN and colorimetric methods, respectively. Data obtained were analyzed in relation to drainage frequency, rice growth stage, and rice yield using the Student's t-test and XLSTAT 7.5.2 statistical software. From the results obtained, the subsurface drainage did not affect significantly SRB populations ($P = 0.187$). However, the drainage affected significantly sulfides concentration in the soil near rice roots ($P = 0.032$). The concentration of sulfides ($P < 0.0001$) in soil near rice roots and the number of SRB ($P < 0.0001$) were significantly higher during the rice tillering and maturity stages. Although no significant difference was observed for rice yield among treatments ($P = 0.209$), the P2 subsurface drainage showed the highest yield and a low concentration of sulfides in soil near rice roots.

Keywords: Sulfate-Reducing Bacteria, Sulfides, Rice, Toxicity, Subsurface Drainage

1. INTRODUCTION

Sulfur is among the most abundant elements on the Earth [1] . In rice soils, S cycle has two major aspects: plant nutrition and sulfides toxicity to the plant. Rice growth requires an adequate supply of S, depending on variety and yield, thus a rice crop removes between 8 and 17 Kg S ha^{-1} from the soil [2] . Microorganisms play an important part in sulphur transformations. Sulfate is taken up as a nutrient and reduced to sulfide, which is then incorporated into sulphur-containing amino acids and enzymes [1] . Sulfate-Reducing Bacteria (SRB) are ubiquitous and quantitatively important members in many ecosystems [1] [3] . They constitute a group of prokaryotes with diverse morphological and metabological characteristics, able to use several low-molecolar organic compounds, including mono- and dicarboxylic aliphatic acids, alcohols, polar aromatic compounds and hydrocarbons [4] - [6] . SRB are involved in numerous reactions under anaerobic environments [7] -[9] . SRB also constitute a diverse group of prokaryotes with the common ability to metabolize oxidized sulfur compounds. They are universally distributed where sulfate reduction is the dominant biomineralization pathway as fens [10] - [12] , estuary sediments [13] - [15] . Rice paddy fields also represent a freshwater environment in which sulfate reduction occurs and from which sulfate reducers have been isolated [6] [12] [16] . SRB gain energy for cell synthesis and growth by coupling the oxidation of these organic compounds or molecular hydrogen

(H_2) to reduction of sulfate (SO_4^{2-}) in sulfide (H_2S, HS^-) [1] [17] . Although most rice soils in the world are S-deficient, significant reduction of sulfate has been reported in a wide range of rice fields [2] [6] . Sulfate reduction is one of the most common and characteristic features shared by various types of paddy soils. It could be responsible to 95% - 97% of sulfides production in rice field [6] . Production of sulfides, at levels that cause rice injury, is observed in many countries, along with Burkina Faso where many rice fields soils are abandoned due to sulfides [18] [19] . Jacq [20] has described sulfides toxicity pathway in rice fields. By toxic sulfides production, SRB may be causative agents of limb discoloration. The high quantity of sulfides ions in the soil solution upsets the mineral element balance in the rice, affects its growth and causes crop yield losses depending on the sulfides concentration in the soil solution and the cultivar tolerance [21] [22] . Hydrogen sulfide may inhibit seed germination [20] , or causes early death of 30% - 100% of crop [23] . Up to now, sulfide toxicity in rice fields had been exclusively attributed to the activity of Sulfate-Reducing Bacteria [24] . The sulfide toxicity generally occurs in degraded paddy soils, poorly drained organic soils, and acid sulfate soils [21] .

For the experiment was developed to measure the effect of subsurface drainage on microbiological and chemical parameters sustaining sulfides toxicity in paddy fields and on rice yield. To eliminate excess sulfides in soil through water management, concrete microplots were constructed, filled with a clay-loam soil and water was drained periodically during rice, as described previously [25] . The Sulfate-Reducing Bacterial (SRB) populations' density and sulfides content in the paddy soil were monitored during the cultural cycle of FKR 19 rice variety.

2. MATERIALS AND METHODS

2.1. Expriment Site Location

The site for the experiments was located at Kamboinse in the central plateau of Burkina Faso (12°26'48"N and 1°33'45"W). Experiments were performed on the experimental site of the International Institute for Water and Environmental Engineering (2iE) from August to December 2011. The physical and chemical properties of the soil used in the experiments were described in our previous report [25] .

2.2. Climatic Characteristics of the Study Site

The site for experiments is located in the central plateau of Burkina Faso which supports five months of rainy season (Juin-October) with peaks from July to August, and seven months of dry season (November-Mai). During the period of study (August-December) thirty rainy days with an average rainfall of 284.8 mm were recorded (Figure 1). The average of daily temperature was 28°C and ranged from 24.6°C (minimum) to 30.9°C (maximum) (Figure 2). The daily relative hygrometry (RH), varied from 39% to 78% (Figure 2), with a sunshine average of 7.26 hours/day (Figure 2).

2.3. Plant Material

For the experiment, FKR 19 rice variety (Farako-Bâ rice) was used. This variety originates from the Asian species Oryza sativa L (Japonica varietal group), and is adapted to the rainfed rice. The variety results from a cross-breeding performed by the Institute for Environment and Agricultural Research (INERA Farako-Bâ, Burkina Faso).With a cycle of seedling-maturity of 120 days, this rice variety has an average yield estimated to 5 - 6 tons/ha.

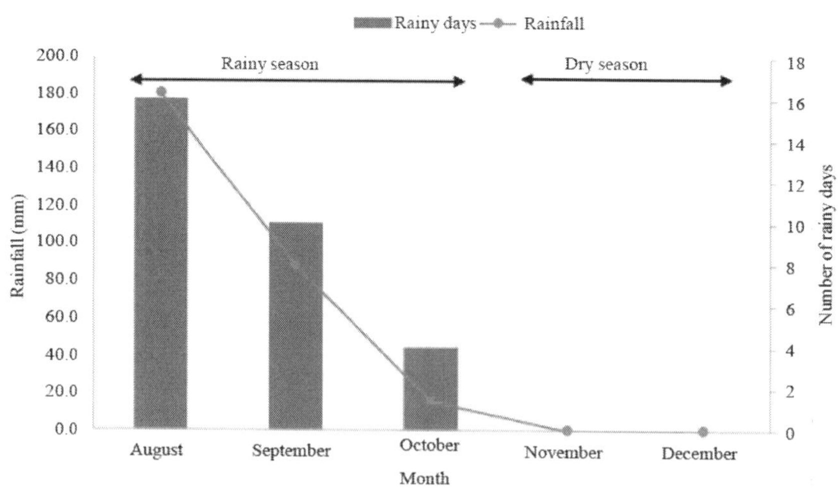

Figure 1. Monthly rainfall and monthly number of rainy days from August to December 2011.

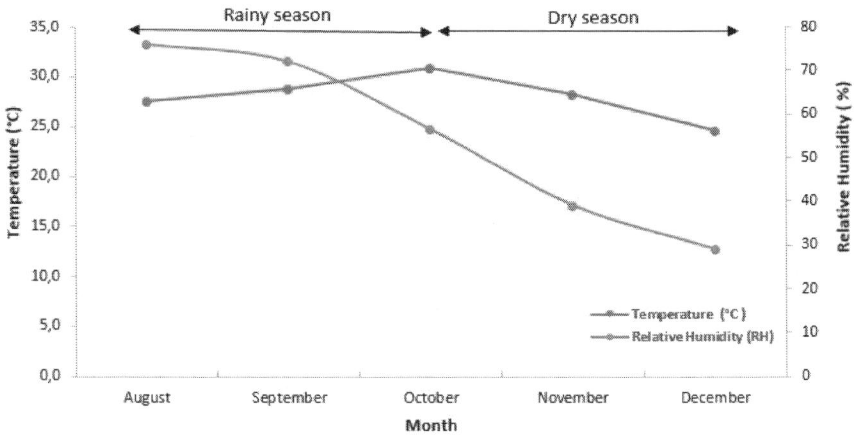

Figure 2. Monthly average temperature and relative humidity from August to December 2011.

2.4. Microplots Experiments

Experiments were carried out as previously described by Otoidobiga et al. [25] and Keïta [26] . Twelve microplots of 1.50 m × 1.20 m × 1.20 m in size (total internal volume: 1.31 m^3) were constructed in reinforced concrete to prevent cracks due to temperature changes following the alternation of the rainy and the dry seasons. At the bottom of each microplot (1.45 m depth), perforated PVC pipes were installed and connected to an external tap through the concrete wall for sub-drain the soil (Figure 3). The design was intended to improve circulation of water and oxygen within the rice roots zone. Microplots were filled with the clay-loam soil described above [25] , and then flooded for 1 week. After the flooding period of soil, two 15-day-old plants of FKR 19 rice were transplanted at 20 cm hill intervals. Then, the plants were continuously flooded from rice transplanting until rice maturity and harvest. Three replications and four modes of drainage were performed throughout the study: without drainage (control: T) and drainage for 7 days (P1), 14 days (P2) and 21 days (P3), respectively by regulating the drained water flow from the bottom tap (Figure 3). Chemical fertilizers (NPK and urea) were applied in the microplots according to the recommended doses of 300 kg/ha for N-P-K (14 - 23 - 14) at transplanting and 100 - 150 kg/ha for urea in two further dressings (at rice transplanting, and 60 days after transplanting, respectively) [27] [28] .

2.5. Enumeration of Sulfate-Reducing Bacterial Populations

Populations of Sulfate-Reducing Bacteria were enumerated in microplots during the rice growth stages by the Most Probable Numbers methods using lactate and sulfate as substrates. The basal medium for enumeration was the medium described by Dianou and Traoré [19] .The medium composition was as follows (per liter of distilled water): [A] sodium lactate 60% (12 ml), K_2HPO_4(0.35 g), KH_2PO_4 (0.25 g), Na_2SO_4 (4.00 g), $MgSO_4$, $7H_2O$ (2.00 g), NaCl (0.5 g), NH_4Cl (2.00 g), yeast extract (1.00 g), resazurin solution 0.1% (1 ml), 1 ml of trace elements solution [29] ; [B] Na_2S, $9H_2O$ (2.5 g), Chlorhydrate of cysteine (1.25 g), $NaHCO_3$ (4.00 g). The heat-stable salts [A] solution was autoclaved and cooled under an atmosphere of N_2. Then, the medium was supplemented with components [B] solution, to final concentration of 2%. The pH was adjusted to 7.2 by addition of NaOH 1N solution. Tubes were incubated at 37°C for two weeks. Formation of colloidal CuS from total sulfides (H_2S, S^{2-} and HS^-) according to Cor-Ruchwich [30] was used for detection of positive tubes after incubation period. The most probable numbers of SRB were calculated from a table of MPN for three tubes.

2.6. Determination of Sulfides Content in Soil

From the soil sampled for bacterial enumeration and at the same periods during the rice cultural cycle, the sulfides were extracted using extraction medium, according to the method adapted from Chaudhry and Cornfield [31] . The extraction medium composition was as follows (per litter of distilled water): $CH_3(COO)_2Zn$, $2H_2O$ (50 g), CH_3COONa, $3H_2O$ (12.5 g). Cor-Ruschwich [30] method was then used to measure the content of sulfides in the soil solution. The extracted sulfides solution (2 ml) was removed by syringe and rapidly injected into tubes containing 2 ml of a copper reagent. Immediately after mixing (for 5s), the absorbance is measured at 480 nm using a Spectronic 61 photometer.

2.7. Statistical Analysis

Data obtained were analyzed for SRB populations' development and activity, drainage mode, rice growth stage and rice yield variations using the Student's t-test and XLSTAT 7.5.2 statistical software. Mean parameters were compared according to the Newman Keuls' test at 5% probability level.

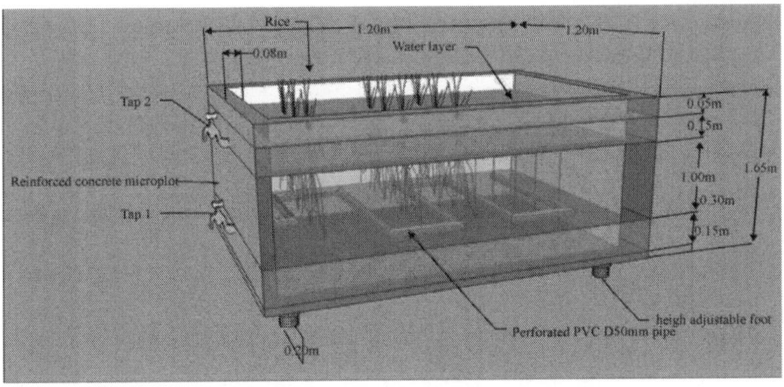

Figure 3. Microplots design for the experiments [25] [26] .

3. RESULTS AND DISCUSSION

3.1. Effect of Soil Moisture on SRB Populations Dynamic and Activity

In the present study, densities of SRB enumerated on soil before flooding, when the soil was dried and at transplanting day after soil flooding, are shown in Figure 4 and expressed as log 10 (most probable population number g^{-1} dry soil).

Sulfides content in soil before flooding when the soil was dried and at transplanting day after one week of soil flooding, are shown in Figure 5 and expressed as log 10 (µg/g dry soil). The concentration of sulfides increased after one week of flooding in the soil for all the microplots.

Rice paddy fields represent a freshwater environment from which sulfate reducers have been isolated and in which sulfate reduction occurs [1] [6] [16] [19] [31] . Many studies showed that sulfate-reducers are common in flooded soils and they are also found near rice roots [8] [32] .

Thus ours results are in agreement with those obtained by Dianou et al. [33] , who reported a population increase of 200%, 139%, 119%, 66%, 40%, 59%, 42% and 22% in the soils of Burkina Faso, after flooding. Garcia et al. [33] reported that SRB populations quantified in the paddy soils of Senegal were influenced by the redox potential related to flooding [34] . Indeed, soon after the flooding of the rice fields, O_2 is removed from the bulk soil via respiratory processes, and large amounts of minerals and nutrients available for bacterial growth were released in the soil solution [16] [35] . In these anaerobiosis conditions the strictly anaerobic bacteria, including sulfate reducers, become active. Thus, significant reduction of sulfate occurs in a wide range of rice fields [16] .

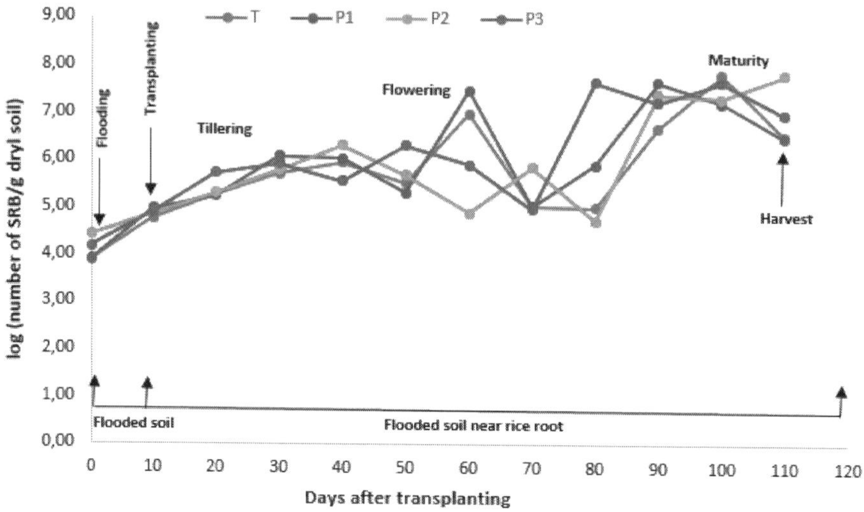

Figure 4. Densities of Sulfate-Reducing Bacteria in soil before flooding, at transplanting day, and in soil near rice roots during the cultural cycle of FKR 19 rice in microplots without drainage (T), and drained for 7 days (P1), 14 days (P2) and 21 days (P3), respectively (means of 3 replicates).

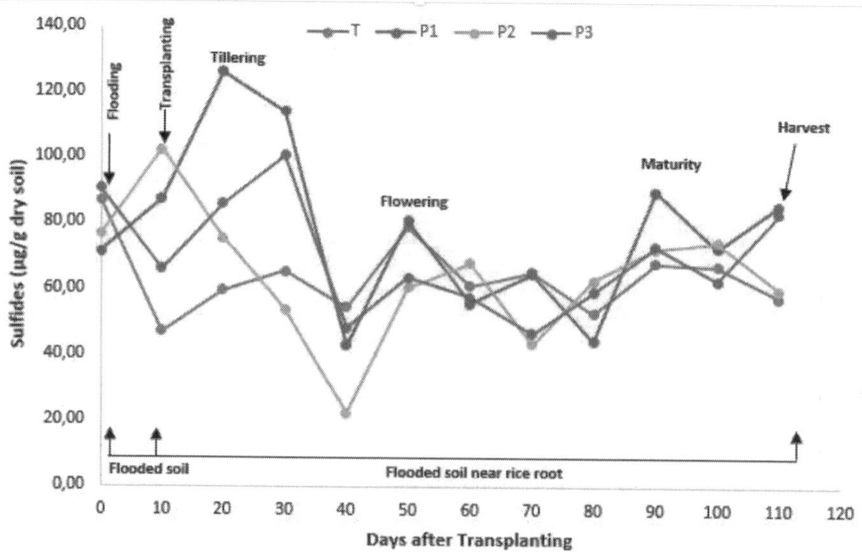

Figure 5. Evolution of the soil sulfides content during the cultural cycle of FKR 19 rice in micro- plots not drained (T), and drained for 7 days (P1), 14 days (P2) and 21 days (P3), respectively (means of 3 replicates).

3.2. Effects of Rice Plant on SRB Populations' Dynamic and Activity

The number of SRB in the soil near rice roots increased after one week of flooding for all microplots (Figure 4). An increase in SRB population number of 95.32%, 93.03%, 86.62% and 88.7% was recorded in the soil of the control (T), P1, P2 and P3 microplots, respectively. Our study also revealed that the number of SRB in soil near rice roots increased gradually with fluctuations from the transplanting day to rice flowering and maturity stages in all the paddy microplots (Figure 4). In most microplots, the highest densities of SRB were recorded from rice tillering and flowering to maturity stages, at which bacterial population number could reach 10^7 to 10^8/g dry soil (Figure 4).

The sulfides content in soil near rice roots also increased gradually with fluctuations during the rice cultural cycle and the highest levels were recorded from rice tillering and flowering to maturity stages at which it could reach 10^2 to 10^3 μg/g dry soil (Figure 5).

Thus, SRB population density and sulfides concentration in soil near rice roots appeared particularly high at the tillering flowering and maturity stages of rice growth (Figure 4 & Figure 5).

Wind et al. [36] who reported that the number of Sulfate-Reducing Bacteria were higher in planted than in unplanted rice microcosms support our findings. Many studies reported also that sulfate reduction rates and sulfate concentration were high at the root surface and near the rice roots [36] [37].

Our results are also in agreement with those of Ouattara [18] and Dianou and Traoré [19] who reported that after rice transplanting the numbers of SRB increased significantly in the plant rhizosphere. Dianou and Traoré [19] reported also that populations increased gradually with fluctuations from transplanting day to rice flowering and maturity stages in the soil near rice roots during paddy plots experiments. Dianou and Traoré [19] and Jacq et al. [38] showed that bacterial population number could reach values of 10^7 to 10^9/g dry soil in most of paddy plot soils of Senegal and Burkina Faso, from rice tilling and flowering to maturity stages, respectively. The rice heading and repining periods which correspond to the highest level of reduced soil conditions in paddy flooded soil, may be conducive to the growth of SRB as underlined Dianou and Traoré [19] . Thus, the high physiological activity of rice plant at these stages may result in the production of more substrates available for bacterial growth [19] [20] [38] .

3.3. Impact of Drainage on SRB Populations' Development and Activity

The variance of the SRB density in soil near rice roots, in relation to drainage and sampling period during rice growth is presented in Table 1. The Newman Keuls' test revealed that the density of SRB in the paddy soil near rice roots was not significantly related to drainage ($P = 0.187$) (Table 1). Moreover, no significant difference was observed between the average numbers of SRB population in the soil of all the microplots (drained or not) (Table 2). However, the number of bacteria in the paddy soil near rice roots was significantly related to sampling period ($P < 0.0001$) and to combined both factors ($P = 0.000$, Table 1). These contrasted results could be partially explained by the MPN method used. Indeed, the analysis of bacterial communities by culturing methods, although valuable to characterize metabolic activities of their members, allows the characterization of only 1% to 3% of the microscopically-detectable cells in soils [39] [40] . Thus, molecular methods could be an indispensable tool to provide a more comprehensive description of the SRB community evolution in relation to drainage [3] [5] [40] [41] .

The variance of the soil sulfides content in relation to drainage and sampling period during rice growth is presented in Table 2. It appeared that sulfides content in the soil near rice roots was significantly related to drainage ($P = 0.032$) and to the sampling period ($P < 0.0001$). SRB have been traditionally considered as strict anaerobes [16] [42] . However, in the present study, the SRB population number in the soil of P1, P2 and P3 drained microplots increased slightly as compared to the control (18.45%, 1.06%, 8.76%, respectively) indicating that some among the SRB can survive and grow in aerobic compartments in the presence of low pressure of oxygen where a surplus of oxygen is released by drainage and by healthy roots [38] [43] .

These results are in agreement with the ones of Johnson et al. [44] and Dolla et al. [45] who found that some species of SRB were capable of slow linear aerobic growth in sulfate-containing medium in the presence of very low concentrations of oxygen. Thus, the abundance and metabolic activity of SRB in oxic zones of numerous biotopes are frequently evaluated as higher than those in neighboring anoxic zones [46] [47] . The high number of SRB found in these oxic environments indicates that these organisms are able to deal with temporarily exposures to elevated oxygen concentration [16] [48] . Wind et al. [36] show also that sulfate reduction also takes place on the roots of intact rice plants when O_2 is allowed to diffuse to the roots through the aerenchyma system of the plants.

The density of SRB populations and the sulfides concentration in drained microplots could also be explained by heterogeneous distribution of oxygen through the drainage system leading to the formation of anoxic compartments [25] . A lack of oxygenation of such microsites in drained microplots may promote the SRB population survival [49] and the sulfides production [50] . Therefore, an efficient oxygenation of soil by water drainage would significantly reduce anaerobic SRB population's development and activity.

Table 1. Variance of SRB number and sulfides content in soil near rice roots in relation to drainage and sampling period during the cultural cycle of FKR19 rice.

Source of variation	DF	Log (SRB number/g dry soil)		Sulfides (µg/g dry soil)	
		F	P	F	P
Drainage	3	1.634	0.187^{ns}	3.062	0.032^{*}
Period	11	27.077	$<0.0001^{**}$	4.195	$<0.0001^{**}$
Drainage x Period	33	2.546	0.000^{**}	1.390	0.110^{ns}

DF = degree of freedom; F = Fisher F; *significant $P < 0.05$; **significant $P < 0.01$; ^{ns}not significant $P < 0.05$.

Table 2. Effect of drainage on SRB population number in soil near rice roots during the cultural cycle of FKR19 rice in microplots not drained (T), and drained for 7 days (P1), 14 days (P2) and 21 days (P3) (means of 3 replicates).

Drainage	Log (SRB number/g dry soil)	Sulfides (µg/g dry soil)
T	5.757^{a}	63.549^{b}
P1	5.936^{a}	77.929^{a}
P2	5.829^{a}	64.234^{b}
P3	6.077^{a}	69.655^{ab}

Means with a same letter within a column are not significantly different according to Newman Keuls' test $P > 0.05$.

3.4. Impact of Drainage on FKR 19 Rice Yield

In our study, symptoms of sulfides toxicity for rice based on IRRI standard evaluation system [51] weren't evidenced during the microplot experiments. Although no significant difference ($P = 0.209$) was found for rice yield among treatments, drained microplots, in particular P2 ones showed the highest yield. Dobermann and Fairhurst [21] reported that an excessive concentration of hydrogen sulfide in the soil results in reduced nutrient uptake due to a decrease of root respiration. The typical symptoms linked to sulfide toxicity involves: rotting of roots, bronzing of leaves, poor growth at the reproductive phase, reduced ability to oxidize iron in the rhizosphere and poor yield [52] - [60] . The symptoms of sulfide toxicity can occur throughout the growth cycle of the rice [21] ; however no critical levels have been established to test sulfide toxicity. Our result can be explained by the chemical fertilizers (NPK and urea) applied in all the microplots. Indeed, according to Dobermann and Fairhurst [21] and Fageria et al. [61] , sulfides toxicity is generally associated with iron toxicity, which is defined as a multiple nutritive disorder further through excesses of H_2S [61] . Deficiencies in minerals may enlarge the root permeability that enhances exudation, oxygen consumption and sulfate reduction, finally leading to severe sulfide toxicity [21] [61] - [63] . Many studies reported that sulfides toxicity for rice depends on nutrient supply; thus, amendment of paddy soil in mineral elements may reduce symptoms of sulfide toxicity for rice [64] [65] .

One raison of our results may be that the FKR 19 rice variety is not susceptible to sulfides toxicity as reported some studies [26] .

Rice plants can develop also physiological avoidance mechanisms to survive under toxic-sulfide condition [36] [49] . Indeed, sulfide toxicity depends on the strength of rice root oxidizing power, H_2S concentration in the soil solution and root health [21] [38] .

Soil oxygenation by water drainage can affect sulfides profiles [49] [66] . The release of oxygen from the roots of rice causes radial redox gradients around the roots, leading to various chemical and microbial oxidation processes in the rhizosphere along with the oxidation of sulfide to sulfate [67] .

4. CONCLUSION

The study revealed important populations of cultivable Sulfate-Reducing Bacteria in soil near roots. The SRB populations and the production of sulfides appeared also impacted mainly by the soil flooding and by the physiological activity of rice plant at tillering, flowering and maturity stages. The effect of subsurface drainage on the above microbial dynamics and activity and the related consequence on rice production highlighted the highest rice yields for drained microplots. Although important concentrations of sulfides were recorded in the drained microplots, no deleterious effect was observed for rice development and rice yield. These findings suggested that the soil oxygenation through subsurface drainage might lead to chemical and microbial oxidation of toxic-sulfides to sulfate.

Thus, an efficient oxygenation of soil by subsurface drainage may be a key factor of sulfides toxicity control in rice paddy field.

ACKNOWLEDGEMENTS

The authors would like to express profound gratitude to International Institute for Water and Environmental Engineering (2iE), CNRST/IRSS, General Direction Meteorology-Burkina Faso, International Foundation for Science, PACER-UEMOA/RABIOTECH, ISP-SUEDE/RABIOTECH, FCN-WAAPP and CNS-FL/WAAPP, for financial and technical supports.

REFERENCES

1. Muyzer, G. and Stams, A.J.M. (2008) The Ecology and Biotechnology of Sulphate-Reducing Bacteria. Nature Reviews Microbiology, 6, 441-456.
2. Freney, L.R., Jacq, V.A. and Baldensperger, L.F. (1982) The Significance of the Biological Sulfur Cycle in Rice Production. In: Dommergues, Y.R. and Diem, H.G., Eds., Microbiology of Tropical Soils and Plant Productivity, Nijhoff/Lunk, Hague, 271-317.
3. Lücker, S., Steger, D., Kjeldsen, K.U., MacGregor, B.J., Wagner, M. and Loy, A. (2007) Improved 16S rRNA-Targeted Probe Set for Analysis of Sulfate-Reducing Bacteria by Fluorescence in Situ Hybridization. Journal of Microbiological Methods, 69, 523-528.
4. Shibata, A., Toyota, K., Miyake, K. and Katayama, A. (2007) Anaerobic Biodegradation of 4-Alkylphenols in a Paddy Soil Microcosm Supplemented with Nitrate. Chemosphere, 68, 2096-2103.
5. Yang, S.Y., Yoshida, N., Baba, D. and Katayama, A. (2008) Anaerobic Biodegradation of Biphenyl in various Paddy Soils and River Sediment. Chemosphere, 71, 328-336.
6. Liu, X.-Z., Zhang, L.-I., Prosser, J.I. and He, J.-Z. (2009) Abundance and Community Structure of Sulfate Reducing Prokaryotes in a Paddy Soil of Southern China under Different Fertilization Regimes. Soil Biology and Biochemistry, 41, 687-694.
7. Jorgensen, B.B. (1982) Mineralization of Organic-Matter in the Sea Bed: The Role of Sulfate Reduction. Nature, 296, 643-645.

8. Gilmour, C.C., Riedel, G.S., Ederington, M.C., Bell, J.T., Benoit, J.M., Gill, G.A. and Stordal, M.C. (1998) Methylmercury Concentrations and Production Rates across a Trophic Gradient in the Northern Everglades. Biogeochemistry, 40, 327-345.

9. Habicht, K.S. and Canfield, D.E. (1996) Sulphur Isotope Fractionation in Modern Microbial Mats and the Evolution of the Sulphur Cycle. Nature, 382, 342-343.

10. Castro, H., Reddy, K.R. and Ogram, A. (2002) Composition and Function of Sulfate-Reducing Prokaryotes in Eutrophic and Pristine Areas of the Florida Everglades. Applied and Environmental Microbiology, 68, 6129-6137.

11. Schmalenberger, A., Drake, H.L. and Küsel, K. (2007) High Unique Diversity of Sulfate-Reducing Prokaryotes Characterized in a Depth Gradient in an Acidic Fen. Environmental Microbiology, 9, 1317-1328.

12. 12. Bao, P., Hu, Z.Y, Wang, X.J., Chen, J., Ba, Y.X., Hua, J., Zhu, C.Y., Zhong, M. and Wu, C.Y. (2012) Dechlorination of p,p'-DDTs Coupled with Sulfate Reduction by Novel Sulfate-Reducing Bacterium Clostridium sp. BXM. Environmental Pollution, 162, 303-310.

13. Kondo, R., Nedwell, D.B., Purdy, K.J. and Silva, S.Q. (2004) Detection and Enumeration of Sulphate-Reducing Bacteria in Estuarine Sediments by Competitive PCR. Geomicrobiology Journal, 21, 145-157.

14. Leloup, J., Quillet, L., Berthe, T. and Petit, F. (2005) Diversity of the dsrAB (Dissimilatory Sulfite Reductase) Gene Sequences Retrieved from Two Contrasting Mudflats of the Seine Estuary, France. FEMS Microbiology Ecology, 55, 230-238.

15. Foti, M., Sorokin, D.Y., Lomans, B., Mussman, M., Zacharova, E.E., Pimenov, N.V., Kuenen, J.G. and Muyzer, G. (2007) Diversity, Activity, and Abundance of Sulfate-Reducing Bacteria in Saline and Hypersaline Soda Lakes. Applied and Environmental Microbiology, 73, 2093-2100.

16. Stubner, S. (2004) Quantification of Gram-Negative Sulphate-Reducing Bacteria in Rice Field Soil by 16S rRNA Gene-Targeted Real-Time PCR. Journal of Microbiological Methods, 57, 219-230.

17. Peck Jr., H.D. (1993) Bioenergetic Strategies of the Sulfate-Reducing Bacteria. In: Odom, J.M. and Singleton Jr., R., Eds., The Sulfate Reducing Bacteria: Contemporary Perspectives, Springer-Verlag, New York, 41-76.

18. Ouattara, A.S. (1992) Contribution to the Study of Iron-Reducing Bacteria and Sulfate in Paddy Soils of the Kou Valley (Burkina Faso). PhD Dissertation, University of Provence Aix-Marseille I, Aix-en-Provence.

19. Dianou, D. and Traoré, S.A. (2000) Sulfate-Reducing Bacterial Populations in Some Lowland Paddy Field Soils of Burkina Faso (West Africa). Microbes and Environments, 15, 41-44.

20. Jacq, V. (1975) The Sulfate-Reduction in Relation with the Root Excretion. Bulletin de la Société Botanique de France, 122, 169-181.

21. Dobermann, A. and Fairhurst, T. (2000) Rice: Nutrient Disorders and Nutrient Management. International Rice Research Institute, Los Ba&ntildeos, 191.

22. Ling, Y.-C., Bush, R., Grice, K., Tulipani, S., Berwick, L. and Moreau, J.W. (2015) Distribution of Iron- and Sulfate-Reducing Bacteria across a Coastal Acid Sulfate Soil (CASS) Environment: Implications for Passive Bioremediation by Tidal Inundation. Frontiers in Microbiology, 6, 1-15.

23. Jacq, V.A. (1977) Rice Susceptibility to Microbial Sulfides. ORSTOM Book, Biology Serial, 12, 97-99.

24. Escoffier, S., Olliver, B., Le Mer, J., Garcin, J. and Roger, P. (1998) Evidence and Quantification of Thiosulfate Reducers Unable to Reduce Sulfate in Rice Field Soils. European Journal of Soil Biology, 34, 69-74.

25. Otoidobiga, C.H., Keita, A., Yacouba, H., Traore, A.S. and Dianou, D. (2015) Dynamics and Activity of Iron-Reducing Bacterial Populations in a West African Rice Paddy Soil under Subsurface Drainage: Case Study of Kamboinse in Burkina Faso. Agricultural Sciences, 6, 860-869.

26. Keïta, A. (2015) Subsurface Drainage of Valley Bottom Irrigated Rice Schemes in Tropical Savannah: Case Studies of Tiefora and Moussodougou in Burkina Faso. PhD Thesis, Wageningen University, Delft.

27. INERA (2000) Descriptive Folders of Rice Cultivars. Institute of Environment and Agricultural Research, Ouagadougou. http://www.inera.bf/

28. Sokona, M.E.B., Boro, A., Hema, A. and Katiella, B. (2010) Diagnostic Study of the Rice Irrigation Scheme of Tiefora, Province of Comoe, Region of the Cascades. Field Report, 2iE, Ouagadougou.

29. Widdel, F. (1983) Methods for Enrichment and Pure Culture Isolation of Filamentous Gliding Sulfate-Reducing Bacteria. Archives of Microbiology, 134, 282-285.

30. Cord-Ruwisch, R. (1985) A Quick Method for the Determination of Dissolved and Precipitated Sulfides in Cultures of Sulfate-Reducing Bacteria. Journal of Microbiological Methods, 4, 33-36.

31. Jacq, V.A. (1978) Adaptation of the Colorimetric Method of Chaudhry and Cornfield to the Dosage of Dissolved Sulfides in a Waterlogged Soil. ORSTOM Book, Biology Serial, 13, 129-132

32. Dianou, D., Lopes, J., Traoré, S.A., Lino, A., Moura, I. and Moura, J.G. (1998) Characterization of Desulfovibrio sp. Isolated from Some Lowland Paddy Field Soils of Burkina Faso. Soil Science and Plant Nutrition, 44, 459-465

33. Garcia, J.L., Rainbault, M., Jacq, V., Rinaudo, G. and Roger, P. (1974) Microbial Activities in Paddy Soils of Senegal: Relations with Physicochemical Characteristics and Influence of the Rhizosphere. Journal of Ecology and Soil Biology, 11, 169-185

34. Loyer, J.Y., Jacq, V.A. and Reynaud, P.A. (1982) Physicochemical Variations in a Flooded Rice Field Soil and Evolution of the Algal Biomass and Microbial Populations in the Sulfur Cycle. ORSTOM Book, Biology Serial, 45, 53-72.

35. Bongoua, D.A.J. (2009) Iron-Reducing Bacterial Communities Implications and Environmental Settings in the Functioning and Rice Fields Soils Quality (Thaïlande and Ivoiry Coast). PhD Thesis, Henri Poincare University, Nancy.

36. Wind, T., Stubner, S. and Conrad, R. (1999) Sulfate-Reducing Bacteria in Rice Field Soil and on Rice Roots. Systematic and Applied Microbiology, 22, 269-279.

37. Wind, T. and Conrad, R. (1997) Localization of Sulfate Reduction in Planted and Unplanted Rice Field Soil. Biogeochemistry, 37, 253-278.

38. Jacq, V.A., Prade, K. and Ottow, L.C.G. (1991) Iron Sulphide Accumulation in the Rhizosphere of Wetland Rice (Oryza sativa L.) as the Result of Microbial Activities. In: Berthelin, J., Ed., Diversity of Environmental Biogeochemistry, Developments in Geochemistry, Vol. 6, Elsevier, Amsterdam, 453-468.

39. Amann, R.L., Ludwing, W. and Schleifer, K.H. (1995) Phylogenetic Identification and in Situ Detection of Individual Microbial Cells without Cultivation. Microbiological Reviews, 59, 143-169.

40. Scavino, A.F., Menes, J., Ferrando, L. and Tarlera, S. (2010) Bacterial Community Analysis of the Water Surface Layer from a Rice-Planted and an Unplanted Flooded Field. Brazilian Journal of Microbiology, 41, 411-419.

41. Ito, T., Nielsen, J.L., Okabe, S., Watanabe, Y. and Nielsen, P.H. (2002) Phylogenetic Identification and Substrate Uptake Patterns of Sulfate-Reducing Bacteria Inhabiting an Oxic-Anoxic Sewer Biofilm Determined by Combining Microautoradiography and Fluorescent in Situ Hybridization. Applied and Environmental Microbiology, 68, 356-364.

42. Postgate, J.R. (1984) The Sulfate-Reducing Bacteria. 2nd Edition, Cambridge University Press, Cambridge, United Kingdom

43. Prade, K. (1987) Influence of Nutrient Supply to the Iron Poisoning of Paddy (O. sativa L.) in the Basse Casamance Senegal. PhD Thesis, University of Hohenheim, Stuttgart. (In German)

44. Johnson, M.S., Zhulin, I.B., Gapuzan, M.-E.R. and Taylor, B.L. (1997) Oxygen Dependent Growth of the Obligate Anaeorobe Desulfovibrio vulgaris Hildenborough. Journal of Bacteriology, 179, 5598-5601.

45. Dolla, A., Fournier, M. and Dermoun, Z. (2006) Oxygen Defense in Sulfate-Reducing Bacteria. Journal of Biotechnology, 126, 87-100.

46. Ravenschalag, K., Sahm, K., Knoblauch, C., Jorgensen, B.B. and Amann, R. (2000) Community Structure, Cellular rRNA Content and Activity of Sulfate Reducing Bacteria in Marine Arctic Sediment. Applied and Environmental Microbiology, 66, 3592-3602.

47. Mussmann, M., Ishii, K., Rabus, R. and Amann, R. (2005) Diversity and Vertical Distribution of Cultured and Uncultured Deltaproteobacteria in an Interdital Mud Flat of the Wadden Sea. Environmental Microbiology, 7, 405-418.

48. Sigalevich, P., Meshorer, E., Helmann, Y. and Cohen, Y. (2000) Transition from the Anaerobic to Aerobic Conditions for the Sulfate-Reducing Bacterium Desulfovidrio oxyclinea Results in Flocculation. Applied and Environmental Microbiology, 66, 5005-5012.

49. Liesack, W., Shnell, S. and Revsbech, N.P. (2000) Microbiology of Flooded Rice Bodies. FEMS Microbiology Reviews, 24, 625-645.

50. Ponnamperuma, F.N. (1972) The Chemistry of Submerged Soils. Advances in Agronomy, 24, 29-96.

51. IRRI (2002) Standard Evaluation System for Rice. International Rice Research Institute, Manila.

52. Vàmos, R. (1959) "Brusone" Disease of Rice in Hungary. Plant and Soil, 11, 65-77.

53. Baba, I. and Iwata, I. (1963) Akiochi Disease of Rice Plants. In: Matsubayashi, M., Ito, R., Nomoto, T., Takase, T. and Yamada, N., Eds., Theory and Practice of Growing Rice, Fuji Publishing Co. Ltd., Tokyo, 149-158.

54. Takijima, Y. (1965) Studies on the Mechanism of Root Damage of Rice Plants in the Paddy Fields (Part 1). Root Damage and Growth Inhibitory Substances Found in the Peaty and Peat Soil. Soil Science and Plant Nutrition, 10, 1-8.

55. Armstrong, J. and Armstrong, W. (2005) Rice: Sulfide-Induced Barriers to Root Radial Oxygen Loss, Fe2+ and Water Uptake, and Lateral Root Emergence. Annals of Botany, 96, 625-638.

56. Mitsui, S. (1965) Dynamic Aspects of Nutrient Uptake. In: Tanaka, A., Ed., The Mineral Nutrition of the Rice Plant, Johns Hopkins University Press, Baltimore, 53-62.

57. Tanaka, A., Ranjit, P., Mulleriyawa, R.P. and Yasu, T. (1968) Possibility of Hydrogen Sulphide Induced Iron Toxicity of the Rice Plant. Soil Science and Plant Nutrition, 14, 1-6.

58. Allam, A.I. and Hollis, J.P. (1972) Sulphide Inhibition of Oxidases in Rice Roots. Phytopathology, 62, 634-639.

59. Joshi, M.M., Ibrahim, I.K.A. and Hollis, J.P. (1975) Hydrogen Sulphide: Effects on the Physiology of Rice Plants and Relation to Straighthead Disease. Phytopathology, 65, 1165-1170.

60. Takijima, Y. (1964) Studies on the Mechanism of Root Damage of Rice Plants in the Peat Paddy Fields. Soil Science and Plant Nutrition, 11, 20-27.

61. Fageria, N.K., Santos, A.B., Barbosa, F.M.P. and Guimar&atildees, C.M. (2008) Iron Toxicity in Lowland Rice. Journal of Plant Nutrition, 31, 1676-1697.

62. Ottow, J.C.G., Prade, K., Bertenbreiter, W. and Jacq, V.A. (1993) Iron Toxicity Mechanisms of Flooded Rice (Oryza sativa L.) in Senegal and Indonesia. In: Raunet, M., Ed., Lowland and Rice-Growing, CIRAD-CA, Montpellier, 231-241.

63. Ethan, S., Odunze, A.C., Abu, S.T. and Iwuafor, E.N.O. (2011) Effect of Water Management and Nitrogen Rates on Iron Concentration and

Yield in Lowland Rice. Agriculture and Biology Journal of North America, 2, 622-629.

64. Muhrizal, S., Shamshuddin, J., Fauziah, I. and Husni, M.H.A. (2003) Alleviation of Aluminum Toxicity in Acid Sulfate Soils in Malaysia Using Organic Materials. Communications in Soil Science and Plant Analysis, 34, 2999-3017.

65. Suswanto, T., Shamshuddin, J., Syed Omar, S.R., Mat, P. and Teh, C.B.S. (2007) Alleviating an Acid Sulfate Soil Cultivated to Rice (Oryza sativa) Using Ground Magnesium Limestone and Organic Fertilizer. Jurnal Tanah dan Lingkungan, 9, 1-9.

66. Hauck, S., Benz, M., Brune, A. and Schink, B. (2001) Ferrous Iron Oxidation by Denitrifying Bacteria in Profundal Sediments of a Deep Lake (Lake Constance). FEMS Microbiology Ecology, 37, 127-134.

67. Wind, T. and Conrad, R. (1995) Sulfur Compounds, Potential Turnover of Sulfate Reduction and Thiosulfate, and Numbers of Sulfate-Reducing Bacteria in Planted and Unplanted Paddy Soil. FEMS Microbiology Ecology, 18, 257-266.

7

CHAPTER

Effects of Potassium Rates and Types on Growth, Leaf Gas Exchange and Biochemical Changes in Rice (Oryza Sativa) Planted Under Cyclic Water Stress

Nurul Amalina Mohd Zain[a,b], Mohd Razi Ismail[b,*]

[a]*Institute of Biological Science, Faculty of Science, University Malaya, Kuala Lumpur, Malaysia*

[b]*Institute of Tropical Agriculture, Universiti Putra Malaysia, 43400 Serdang, Selangor Darul Ehsan Malaysia*

ABSTRACT

Three levels of potassium rates [80 kg K_2O ha^{-1}, 120 kg K_2O ha^{-1} and 160 kg K_2O ha^{-1}) and two types potassium (KCl and K_2SO_4) on rice under cyclic water stress 15 days and the absolute control (80 kg K_2O ha^{-1} of KCl fertilizer on rice under control flooded) were exposed to rice to investigate the influence of potassium in minimizing cyclic water stress effects in rice. It was found as fertilization rates increased from 80 > 120 > 160 kg K_2O ha^{-1} the production of proline was increased. The increase in proline production was simultaneously enhanced the production of Catalase and Maliondialdehyde. As the potassium rate increased from 80 > 120 > 160 kg K_2O ha^{-1}, the transpiration rate was observed to be increased in both potassium types. The result suggested that high potassium rates would reduce water stress effects by having high transpiration rate. High transpiration rate would increase the nutrient uptake that would repair the damage tissue under water stress thus reduce the oxidative stress of rice under water stress condition. This been showed by high significant positive correlations of transpiration rate with CAT activity ($r = 0.871; p \leq 0.05$), MDA ($r = 0.914; p \leq 0.05$) and Proline ($r = 0.842; p \leq 0.05$). It was found that the increase of K rates by KCl increased NAR higher than the increased K rates in K_2SO_4. This might be due to higher absorption of K element in rice by KCl compared to K_2SO_4. The study has showed that application of potassium either KCl or K_2SO_4 would minimize the effects on rice growth and physiology under cyclic water stress. The current study also suggested that plant tolerate to cyclic water stress by increased the production of proline, MDA and decrease of Catalase activity to protect the plant from damage from water stress.

Keywords: Cyclic water stress; Potassium; Growth and yield; Leaf gas exchange responses;Biochemical changes

1. INTRODUCTION

Rice is the major source of carbohydrate to millions of people world over, particularly in Asia (Price and Tomos, 1997). Rice productivity has been reduced due to global warming that has limited the availability of fresh water for rice crop, particularly under rain-fed conditions (Liu et al., 2010). To produce 1 kg of grain, farmers have to supply 2–3 times more water in rice fields than other cereals (Banoc et al., 2000). In Asia, more than 80% of the developed freshwater resources are used for irrigation purposes; about half of which is used for rice production (Asch et al., 2005). Among all the environmental stresses, drought is one of the most severe stresses for rice growth and productivity. Drought stress reduces both nutrient uptake by the roots and transport from roots to the shoots, due to restricted transpiration rates and impaired active transport and membrane permeability (Colmer et al., 1998). Rice is very sensitive to water stress and attempts to reduce water inputs may tax true yield potential (Fukai and Cooper, 1995). The challenge is to develop novel technologies and production systems that would allow rice production to be maintained at the face of declining water availability.

The application of potassium, K can minimize the drought effects of rice. Potassium has substantial effect on enzyme activation, protein synthesis, photosynthesis, stomatal movement and water relations (turgor regulation and osmotic adjustment) in plants (Marschner 1995). Increased application of K has been shown to enhance photosynthetic rate, plant growth, yield, and drought resistance in different crops under water stress conditions (Yadav et al., 1999, Egila et al., 2001 and Pervez et al., 2004). K-fed plants maintained higher leaf water potential, turgor potential and relative water content and lower osmotic potential as compared to untreated plants of *Vignaradiata* (Nandwal et al., 1998), maize (Premachandra et al., 1991), and wheat (Pier and Berkowitz, 1987 and Sen Gupta et al., 1989) grown under water stress. Nodulation, nitrogenase activity and dry matter yield increased with incremental K supply in broad bean grown at moisture level of only 1/4 of field capacity (Abd-Alla and Wahab, 1995). K is predominant in accumulating solute during drought in tropical grasses (Ford and Wilson, 1981), soybean (Itoh and Kumara, 1987), maize (Premachandra et al., 1991), cotton (Pervez et al., 2004) and olive trees and sunflower (Benlloch-Gonzalez et al., 2008) and significantly contributed to osmotic adjustment.

Under water-deficit conditions, K nutrition increases crop tolerance to water stress by utilizing the soil moisture more efficiently than in K-deficient plants. The positive effects of K on water stress tolerance may be through promotion of root growth accompanied by a greater uptake of nutrients and water by plants (Rama Rao, 1986) and through the reduction of transpirational water loss (Beringer and Trolldenier, 1978). Also, K maintains the osmotic potential and turgor of the cells (Hsio, 1973 and Lindhauer, 1985) and regulates the stomatal functioning under water stress conditions (Umar, 2006 and Kant and Kafkafi, 2002), which is reflected in improved crop yield in drought conditions (Umar and Bansal, 1997 and Umar and Moinuddin, 2002). Besides, it takes part in many essential processes in plants (Marschner, 1995) and enhances photosynthetic rate, plant growth and yield under stress conditions (Egila et al., 2001, Sharma et al., 1996, Tiwari et al., 1998 and Umar and Moinuddin, 2002). The protective role of K in plants suffering from drought stress has been attributed to the maintenance of a high pH in stroma and against the photo-oxidative damage to chloroplasts (Cakmak, 1997).

Despite can alleviate water stress effect on rice, potassium are an essential component of plant nutrition and usually used in the form of chloride (KCl), sulfate (K_2SO_4) and nitrate (KNO_3). However, in Malaysia two types potassium are usually found (KCl and K_2SO_4). The used of KCl and K_2SO_4 in reducing the water stress effects was never reported. There was also no recommendation rate of what rates are efficient to tolerate water stress. With that, the objective of present study was (1) to identify the best types (KCl and K_2SO_4) and rates of potassium application (80, 120 and 160 kg K_2O ha^{-1}) in alleviation of water stress in rice. The project was also (2) to investigate the growth and yield, leaf gas exchange and biochemical of rice under cyclic water stress and potassium fertilization. The last objective was (3) to understand the relationship between the yield and leaf gas exchange properties in alleviating water stress effects with potassium fertilization. Understanding the physiological and morphological responses of rice under different cyclic water stress and potassium fertilization is imperative for efficient management of agronomical inputs (irrigation and nutrient) because two of it is important in rice cultivation.

2. MATERIALS AND METHODS

2.1. Experimental location, plant materials and treatments

MR220 rice seeds provided by MARDI Genebank Seberang Perai were germinated and transplanted to tank containing Soil Bakau Series (EC = 2.83 dS m^{-1}; pH 5.1) on 15 days after sown (DAS) in the greenhouse of Ladang 2, Universiti Putra Malaysia, Serdang, Selangor, Malaysia (3°0′34.9020″N latitude; 101°42′7.0416″E longitude; altitude of 46 m above sea level) from December 2009 until April 2010. The light intensity inside the glasshouse was from 225 to 1450 μmol m^{-2} s^{-1}. The relative humidity was between 50 and 60% with a mean atmospheric pressure of 1.013 kPa. The tank with size approximately (55 × 55 × 55) cm had 9 seedlings transplanted in 3 rows × 3 column with space of 20 cm. From 0 DAS until 30 DAS, seedlings were watered with adequate amount of water for 1 cm water level from soil. And on 30 DAS, all tanks were watered at 10 cm water level from soil before first treatment was applied. This factorial experiment was arranged in randomized complete block design (RCBD) replicated four times with two factors i.e. potassium types (KCl and K_2SO_4) and potassium rates in K_2O (80 kg K_2O/ha (control), 120 kg K_2O/ha and 160 kg K_2O/ha). The rice seedlings were exposed to cyclic water stress for 15 days. In this experiment, absolute control was rice irrigated at control flooded + 80 kg K_2O/ha of KCl fertilizer based on standard grower practices in Malaysia. Other fertilizer important like urea (46% N) and triose super phosphate (46% P_2O_5) being added as normal fertilization applied at 120 kg N/ha and 70 kg P_2O_5/ha respectively.

2.2. Measurement

2.2.1. Growth and yield

Rice was harvested when 70% of rice shown ripening colour. Rice panicles with spikelets were cut and collected for every unit before being dried in oven at 60 °C for 2 days. Total panicles for every unit were counted. Then, rice spikelets or grains were separated from the panicle into 2 groups, unfilled grain and filled grain. From here, grain yield, 1000-grains weight and filled spikelets percentage were counted. The rice yield per pot was measured based on grain of 14% moisture content yield weight per pot. Meanwhile for straw biomass was collected from the remaining of plant (without panicle and root) and was dried in a large oven at 70 °C for 3 days. Then the straw was measured for average of 3 hills per unit and being calculated for pot area. Meanwhile harvest index was calculated based on ratio of economic yield to total biomass produced. Water productivity in terms of irrigation was determine by following Molden et al. (2003) as yield (kg ha^{-1})/irrigation inflow (L ha^{-1}). Total plant biomass was taken by calculating the dry weight of root, boles and leaves per seedling. Destructive plant analysis was carried out every 3 weeks for 15 weeks. The plant parts were placed in paper bags and oven dried at 80 °C until constant weight was reached using electronic weighing scale (CDS 125, Mitutoyo Inc., Japan). Leaf area per plant was measured using a leaf area meter (LI-3100, Lincoln Inc., USA). The leaves were arranged within the field of view, and overlapping of adjacent leaves was avoided. Growth analysis was calculated on an individual plant basis through the measurement of total plant leaf area and dry weight. Net assimilation rate (NAR) was calculated based on biomass and leaf area parameters according to the formula reported by Khan et al. (2012).

2.2.2. Leaf gas exchange measurements

The measurement was obtained from a closed infra-red gas analyzer LICOR 6400 Portable Photosynthesis System (IRGA, Licor Inc., Lincoln, NE, USA). The measurements used optimal conditions set of 400 μmol mol^{-1} CO_2, 30 °C cuvette temperature, 60% relative humidity with air flow rate set at 500 cm^3 min^{-1}, and modified cuvette conditions of 225, 500, 625 and 900 μmol m^{-2} respectively, photosynthetically photon flux density (PPFD), according to the irradiance treatment. Gas exchange measurements were carried out between 0900 h and 1100 h, using fully expanded young leaves numbered 3 and 4 from plant apex to record net photosynthesis rate (A). The operation was automatic and the data were stored in the LI-6400 console and analyzed by "Photosyn Assistant" software (version 1.0; Dundee Scientific: Dundee, Scotland, UK, 2000). Several precautions were taken to avoid errors during measurements. Leaf surfaces were cleaned and dried using tissue paper before being enclosed in the leaf cuvette. All measurements were record on fully expanded leaf. Water use efficiency was measured based on the equation by Tanner and Sinchlair (1983):

$WUE = \frac{P_n}{E_p}$ P_n = net Co2 assimilation rate (μmol/m^2/s) and E_p is evapotranspiration rate (mmol/m^2/s).

2.2.3. Assay of Catalase (CAT) activity

2.2.3.1. Preparation of enzyme extracts

To determine the enzymatic activities of the antioxidant proteins, a crude enzyme extracts was prepared by homogenizing 500 mg of leaf tissue in extraction buffer containing 0.5% Triton X-100 and 1% polyvinylpyrrolidone in 100 mM potassium phosphate buffer (pH 7.0) using a chilled mortar and pestle. The homogenate was centrifuged at 15,000 rpm for 20 min at 4 °C. The supernatant was used for the enzymatic assays. Catalase activity (CAT; EC 1.11.1.6) was determined by consumption of H_2O_2using the method of Aebi (1983). The reaction mixture (3 mL) contained 50 mM potassium phosphate buffer pH 7.0, 15 mM H_2O_2 and 50 μL enzyme extract. The reaction was initiated by adding the H_2O_2. The consumption of H_2O_2 was monitored spectrophotometrically at 240 nm for 3 min. Enzyme activity was expressed in micromole per liter H_2O_2 min^{-1}.

2.2.4. Malondialdehyde (MDA content)

1 g of fresh leaves was ground in 5 mL of 0.5% trichloroacetic acid (TCA) and centrifuged at 1000× g for 15 min. The mixture containing 1 mL of the supernatant and 4 mL of 0.5% thiobarbituricacid (TBA) in 20% TCA was heated at 100 °C for 30 min and then cooled to room temperature. The specific absorbance (at 532 nm) of the extract (relative to the background absorbance at 600 nm) was determined. The concentration of malondialdehyde (MDA) was expressed in μmol g^{-1} FW (fresh weight of leaves), using a molar extinction coefficient equal to 155 × 105 $mmol^{-1}$ cm^{-1}.

2.2.5. Proline determination

About 5 mg of lyophilized and homogenized samples were extracted in 0.5 mL of 3% 5-sulphosalicylic acid for 15 min. Then the samples were centrifuged at 21,000 × *g* for 15 min. The clear supernatant (200 μL) was transferred to polypropylene screw cap vials, after which 200 μL of concentrated formic acid and 400 μL of 3% ninhydrin reagent in 2-methoxyethanol were added. Samples were heated for 0.5 h at 100 °C in a water-bath, and then transferred to 96-well plates. Absorbance was measured at 514 nm on a micro-plate reader (Synergy 2, Bio-Tek, Winooski, VT, USA). The level of proline was expressed in μg g^{-1} DW (dry weight of leaves).

2.3. Statistical analysis

Data were analyzed using Proc GLM using SAS 9.2 (32). Mean separation test between treatments was performed using Least Significant Different, LSD and standard error of differences between means was calculated with the assumption that data were normally distributed and equally replicated.

3. RESULT AND DISCUSSION

Overall results showed that half of parameters were influenced by interaction effects of potassium rates and types. There were shoot dry weight, leaf area, total spikelet panicle^{-1}, total empty spikelet panicle^{-1}, biomass rate, net assimilation rate, transpiration rate and water use efficiency of rice. Based on results, potassium types only had significant different on total tillers plant^{-1}, leaves number plant^{-1} and 1000 grains weight of rice. This showed that the companion nutrient (chloride and sulphate) in potassium fertilizer had only affected on tiller and leaves growth and grains weight. Meanwhile potassium rate had more influenced on catalase activity, free proline, Maliondialdehyde content, panicle dry weight hill^{-1}, root dry weight hill^{-1}, rice yield and harvest index.

3.1. Panicle dry weight hill^{-1}, root dry weight hill^{-1}, rice yield and harvest index

It was observed that panicle dry weight hill^{-1}, root dry weight hill^{-1}, rice yield and harvest index was influenced by potassium rates ($P < 0.05$; Table 1). The panicle dry weight hill^{-1}was highest in control that record 23.94 followed by 120 kg K_2O ha^{-1} (21.62), 160 kg K_2O ha^{-1} (18.29) and 80 kg K_2O ha^{-1} (12.31). There was no significant in root dry weight hill^{-1} in control, 80 kg K_2O ha^{-1} and 120 kg K_2O ha^{-1} potassium application compared to the application of 160 kg K_2O ha^{-1} that gave the highest root dry weight (0.45 g). Table 1 had showed that rice yield was lowest under 80 kg K_2O ha^{-1} that record 129.43 g tank^{-1} compared to control (319.15 g tank^{-1}), followed by 120 kg K_2O ha^{-1}(211.40 g tank^{-1}) and 160 kg K_2O ha^{-1} (191.19 g tank^{-1}). The result suggested that increasing of potassium fertilizer rate during cyclic water stress 15 days reduced the yield difference to control. The same attributes to harvest index when the highest harvest index was obtained in control (0.52), followed by 120 (0.42), 160 (0.40) and 80 kg K_2O ha^{-1} (0.30). Rice yield and harvest index had a significant correlation with leaf area ($r = 0.921$; $p \leq 0.05$; Table 2). This included more leaf area needed by this variety to obtained higher yield and harvest index. This indicates as the rate of potassium take increases, there will be increases of leaf area that simultaneously increased the yield and harvest index of rice. The same result was observed as the rate of potassium increase in Sorghum the yield and harvest index was significantly increased (Shahid, 2006).

Table 1. Impact of potassium rates on growth and yield of rice.

Potassium rates ($kg\ K_2O\ ha^{-1}$)	Panicle dry weight $plant^{-1}$ (g)	Root dry weight $plant^{-1}$ (g)	Rice yield ($g\ tank^{-1}$)	Harvest index (HI)
Control*	23.94 ± 3.23a	0.34 ± 0.02b	319.15 ± 22.97a	0.52 ± 0.12a
80	12.31 ± 1.25d	0.31 ± 0.12b	129.43 ± 25.81c	0.30 ± 0.22d
120	21.62 ± 0.65b	0.31 ± 0.08b	211.40 ± 23.26b	0.42 ± 0.42b
160	18.29 ± 1.27c	0.45 ± 0.06a	191.19 ± 44.48b	0.40 ± 0.31c
Main effects				
K Rates	**	**	**	**
K Types	ns	ns	ns	ns
K R × K T	ns	ns	ns	ns

All analyses are mean ± standard error of means (SEM), *control = 80 kg $K_2O\ ha^{-1}$ KCl fertilizer + control flooded, ** = significant at 1% level, ns = no significant ($P < 0.05$).

Table 2. Pearsons correlations between all parameters measured in the study.

Potassium rates (kg K_2O ha^{-1})	Panicle dry weight $plant^{-1}$ (g)	Root dry weight $plant^{-1}$ (g)	Rice yield (g $tank^{-1}$)	Harvest index (HI)
Control*	23.94 ± 3.23a	0.34 ± 0.02b	319.15 ± 22.97a	0.52 ± 0.12a
80	12.31 ± 1.25d	0.31 ± 0.12b	129.43 ± 25.81c	0.30 ± 0.22d
120	21.62 ± 0.65b	0.31 ± 0.08b	211.40 ± 23.26b	0.42 ± 0.42b
160	18.29 ± 1.27c	0.45 ± 0.06a	191.19 ± 44.48b	0.40 ± 0.31c
Main effects				
K Rates	**	**	**	**
K Types	ns	ns	ns	ns
K R × K T	ns	ns	ns	ns

All analyses are mean ± standard error of means (SEM), *control = 80 kg K_2O ha^{-1} KCl fertilizer + control flooded, ** = significant at 1% level, ns = no significant ($P<0.05$).

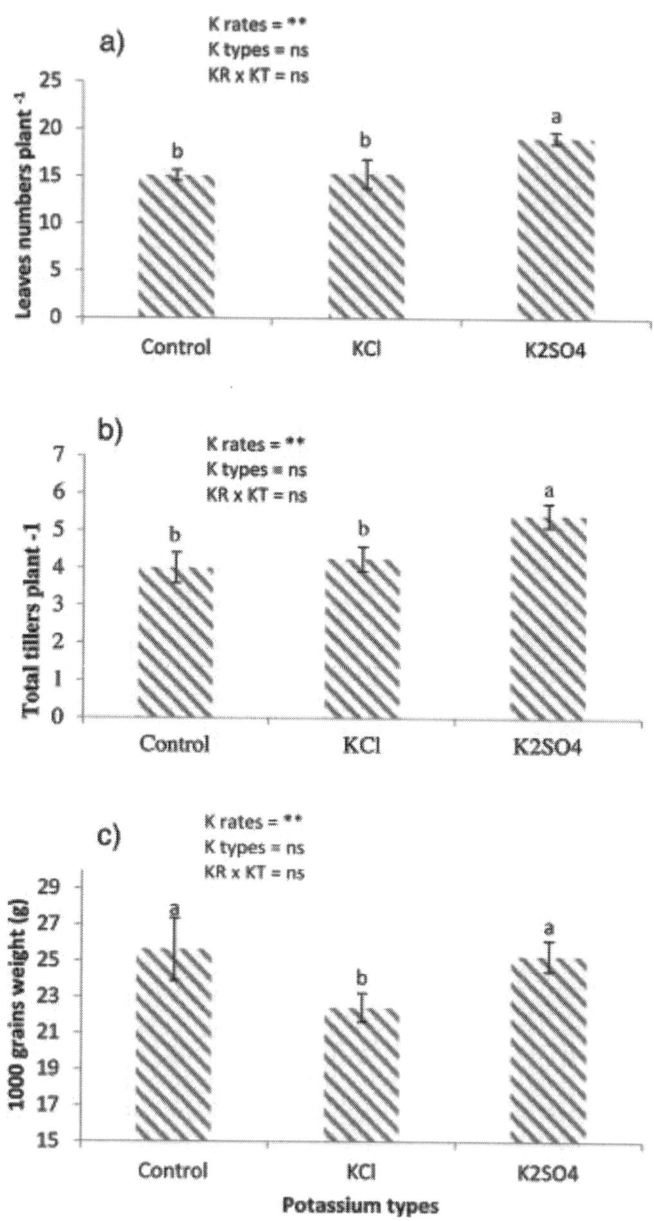

Figure 1. Impact of potassium types on leaves numbers plant^{-1} (a), total tillers plant^{-1} (b) and 1000 grains weight (c) of rice. $N = 28$. Bars represent standard error of differences between means. Control = 80 kg K_2O ha^{-1} KCl + control flooded, ** = significant at 1% level, ns = no significant ($p < 0.05$).

3.2. Leaves number, total tillers and 1000 grains weight

The leaves number, total tillers and 1000 grains weight was influenced by the potassium types ($p \leq 0.05$; Fig. 1). The leaves number was statistically high in K_2SO_4 (17.14) but control (14.89) had no significant different with KCl (14.80) application. In total tillers, the application of K_2SO_4 produced highest tillers number (5.57) compared to KCl (4.17) and control (4.02). Both result on leaves number and total tillers showed that K_2SO_4produced highest level greater than control. There were no significant different in 1000 grains weight between control (25.01) and K_2SO_4 (25.14), however KCl record the lowest 1000 grains weight (23.12). The increase in the vegetative growth of rice by application of K_2SO_4 was observed by Hara and Khan et al. (2012). The leaves number, total tillers and 1000 grains weight have a significant positive correlation with rice yield (Leaves number, $r = 0.92$, tillers, $r = 0.87$ and 1000 grains weight, $r = 0.97$; $p \leq 0.05$; Table 2). The data indicates that the increases in the leaves number, total tillers and 1000 grains weight were simultaneously increase the rice yield. This showed the importance of these components influence the rice yield.

3.3. Leaf area and shoot dry weight

The leaf area and shoot dry weight was influenced by interaction effects between potassium rate and types ($p \leq 0.05$; Fig. 2). The leaf area and shoot dry weight showed almost the same trends where the highest leaf area was in shoot dry weight was in K_2SO_4 + 120 kg K_2O ha^{-1} that recorded 181.35 cm^2 and 1.64 g respectively. The lowest in shoot dry weight was observed in K_2SO_4 + 80 kg K_2O ha^{-1} that record 0.66 g meanwhile the highest was showed in K_2SO_4 + 120 kg K_2O ha^{-1} that 2 fold greater than control respectively. And K_2SO_4 + 120 kg K_2O ha^{-1}also had highest level followed by KCl + 80 kg K_2O ha^{-1}. The current study suggested that application of K_2SO_4 at 120 kg K_2O ha^{-1} can enhance the leaf area and shoot biomass in rice. High significant correlation between leaf area and shoot dry weight ($r = 0.98$; $p \leq 0.05$; Table 2) suggested that a lot of energy from light reaction being invested to the above ground biomass production.

3.4. Total spikelet panicle^{-1} and empty spikelet panicle^{-1}

The interaction between potassium rates and types has influenced total spikelet panicle^{-1} and empty spikelet panicle^{-1} ($p \leq 0.05$; Fig. 3). Total spikelet panicle^{-1} in KCl was highest in application of 120 kg K_2O ha^{-1} (214.54) and at K_2SO_4 + 80 kg K_2O ha^{-1} the total spikelet/panicle was optimized (193.66). The application of KCl + 80 kg K_2O ha^{-1}(168.25) record the lowest total spikelet/panicle compared to the control (186.01) with the difference was only 9.5%. Interestingly, it was observed that the empty spikelet panicle^{-1}was enhanced in all combination treatments compared to the control. The highest empty spikelet panicle^{-1} was observed K_2SO_4 + 80 kg K_2O ha^{-1} (112.33) and lowest in application of K_2SO_4 + 120 kg K_2O ha^{-1} that recorded 66.16. The result showed that increasing potassium rate reduce percentage of empty spikelet, and the present study showed that reduction in yield of rice under influence of water stress in rice might be contributed to increase in empty spikelet panicle^{-1}. The increased in incidence of empty spikelet panicle^{-1} under water stress was never reported, however study by Rad et al. (2012) and Aref and Rad (2012) have showed that increased in salinity stress have increased the incidence of empty spikelet panicle^{-1} in rice. This concludes that drought stress might reduce the rice yield by increasing the increase in empty spikelet panicle^{-1}.

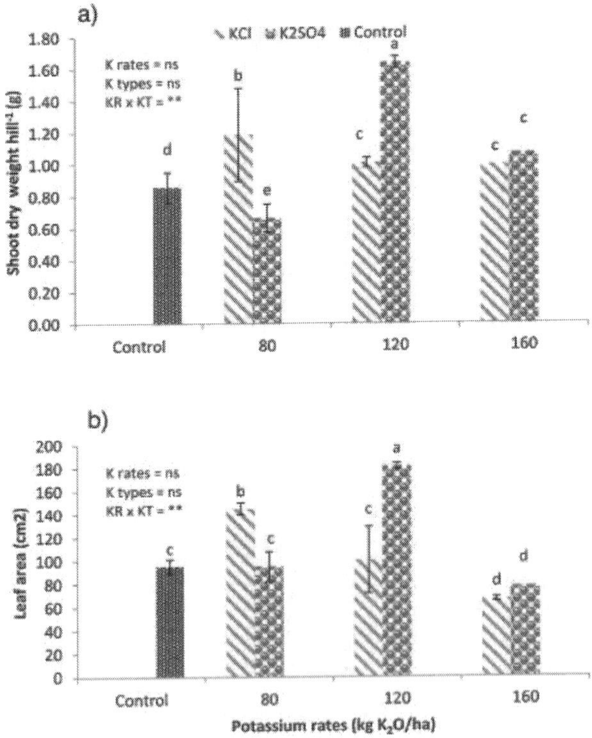

Figure 2. Interaction effects between potassium rates and sources on shoot dry weight hill^{-1} (a) and leaf area (b) of rice. $N = 28$. Bars represent standard error of differences between means. Control = 80 kg K_2O ha^{-1}KCl + control flooded, ** = significant at 1% level, ns = no significant ($p < 0.05$).

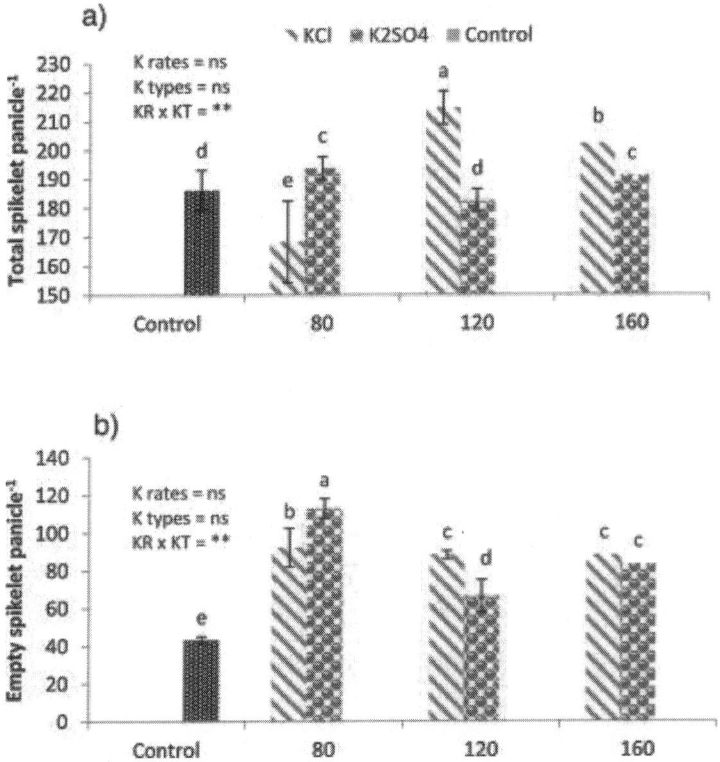

Figure 3. Interaction effects between potassium rates and sources on (a) total spikelet panicle^{-1} and (b) total empty spikelet panicle^{-1}. $N = 28$. Bars represent standard error of differences between means. Control = 80 kg K_2O ha^{-1} KCl + control flooded, ** = significant at 1% level, ns = no significant ($p < 0.05$).

3.5. Net assimilation rate

Interaction effects between potassium rate and types have influenced the net assimilation rate (NAR) ($p \leq 0.05$; Fig. 4). The highest NAR was recorded at KCl + 160 kg K_2O ha^{-1} that recorded 0.029 g cm^{-2} week^{-1}, while the lowest at K_2SO_4 + 80 kg K_2O ha^{-1} that just recorded 0.013 g cm^{-2} week^{-1}. The highest NAR under application K_2SO_4 was at 120 kg K_2O ha^{-1} (0.022 g cm^{-2} week^{-1}). This indicated that addition KCl can enhance NAR better than addition of K_2SO_4. The higher NAR with KCl compared to K_2SO_4 might be due to higher absorption of K element in rice leaves by KCl compared to K_2SO_4. This was in agreement with the findings of Khadr et al. (2004) that observed that application of KCl improved the K uptake in sugar cane compared to the use of K_2SO_4 by 3%. This suggested that K was more available to the plant in KCl compared to K_2SO_4. Correlation table showed that NAR have a significant positive correlation with transpiration ($r = 0.89$; $p \leq 0.05$; Table 2) and rice yield ($r = 0.912$;$p \leq 0.05$; Table 2). This suggests that increases in transpiration rate under high application of potassium might enhanced photosynthesis rate and simultaneously enhanced the rice yield. High significant positive correlation was also observed between NAR and leaf area ($r = 0.92$; $p \leq 0.05$; Table 2) indicating enhancement of growth of rice are mostly influenced by increase in light absorption area compared to the other growth factors. Similar result was also obtained by Zeng et al. (2010) in rice where they observed linear correlation between NAR and leaf area.

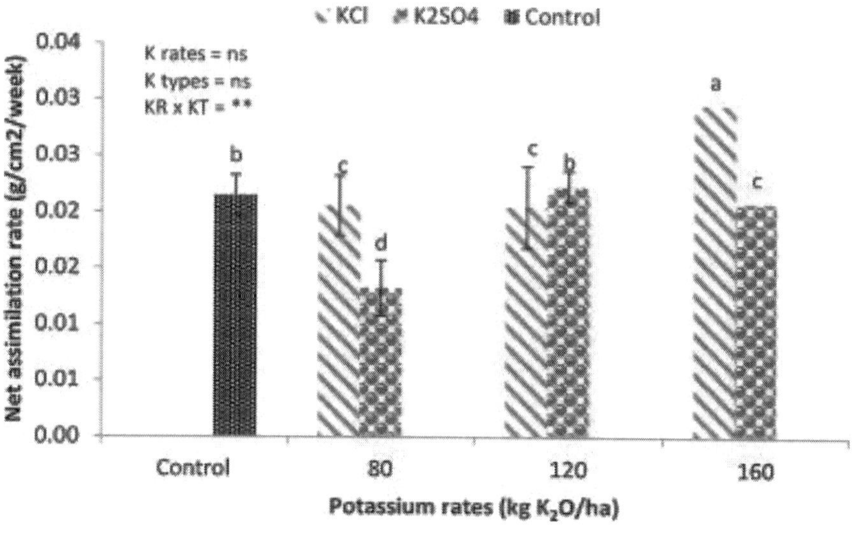

Figure 4. Interaction effects between potassium rates and sources on net assimilation rates of rice. $N = 28$. Bars represent standard error of differences between means. Control = 80 kg K_2O ha^{-1} KCl + control flooded, ** = significant at 1% level, ns = no significant ($p < 0.05$).

3.6. Catalase activity

From the present study it was observed that CAT activity was influenced by potassium rate ($p \leq 0.05$; Fig. 5). The CAT activity was reduced to 40% and 12% respectively in 160 kg K_2O ha^{-1} and 120 kg K_2O ha^{-1} potassium rate compared to the control however, application of 80 kg K_2O ha^{-1} potassium increased the CAT activity 22% higher than the control treatment. The data indicate that application of high potassium would reduce the CAT activities. Catalase functions as effective quenchers for ROS (Foyer et al., 1994). CAT plays an essential role in scavenging from H_2O_2 toxicity. This showed that oxidative stress can be reduced by high application of potassium in rice.

These result also implied that fertilization with high potassium can play important protective role in H_2O_2scavenging processes. It was reported by Tripathi et al. (2009) that proteins such as thioredoxin, glutaredoxin and cyclophilin are known to facilitate the regeneration of the reduced (catalytically active) form of peroxyredoxin that plays an important role in reducing the ROS formation in plants under biotic and abiotic stress, the increase application of potassium fertilizer would increase these proteins (Chaitanya et al., 2002). The present study was in agreement with Ibrahim and Jaafar (2012) on *Labisia pumila*where application of potassium from 90 > 270 kg ha^{-1} have reduced the CAT activity and indicate low oxidative stress of the plants. It also was observed from the correlation table that CAT have a significant negative relationship with transpiration rate ($r = -0.871; p \leq 0.05$; Table 2) that indicate reduction in oxidative stress might be due to increase in transpiration rate under high potassium fertilization, this suggest that enhancement of transpiration rate might reduce the water stress effects in rice.

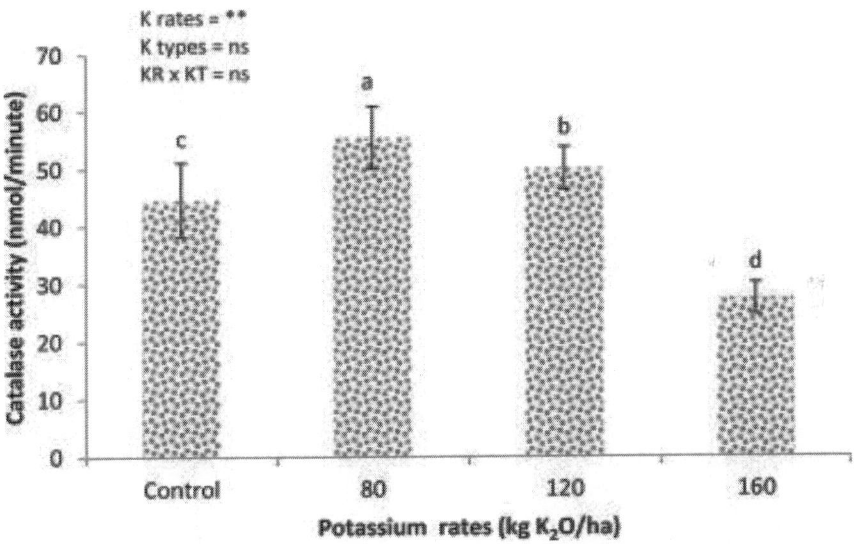

Figure 5. Impact of potassium rates on catalase activity of rice. $N = 28$. Bars represent standard error of differences between means. Control = 80 kg K_2O ha^{-1} KCl + control flooded, ** = significant at 1% level, ns = no significant($p < 0.05$).

3.7. Proline and MDA

It was observed that potassium levels have influenced the proline and lipid peroxidation of the rice ($p \leq 0.05$; Fig. 6). Generally as potassium levels increased, the production of proline has shown to increased. Proline content was enhanced by 152, 257 and 338% as compared to control that just recorded 2.1 mg g^{-1} fresh weight. The lipid peroxidation was highest under 160 kg K_2O ha^{-1} that recorded 6.4 mg g^{-1} followed by 120 kg K_2O ha^{-1} (5.2 mg g^{-1}), 80 kg K_2O ha^{-1} (4.2 mg g^{-1}) and Control at 3.1 mg g^{-1} (Fig. 6). In Fig. 7 it was shown that the increased in proline content have increased the production in catalase and lipid peroxidation in rice, where proline have a significant positive regression with Catalase ($r = 0.910$; $p \leq 0.05$) and lipid peroxidation ($r = 0.715$; $p \leq 0.05$;Table 2). The same observation was also obtained in *Vacia faba* (Siddiqui et al., 2012) and *Halocnemum strobilaceum* (Su et al., 2012) under cadmium and salt stress. This result indicates that under application of high potassium to rice, stress can occurred and be manisfested by steady increase in proline, and maliondialdehyde content.

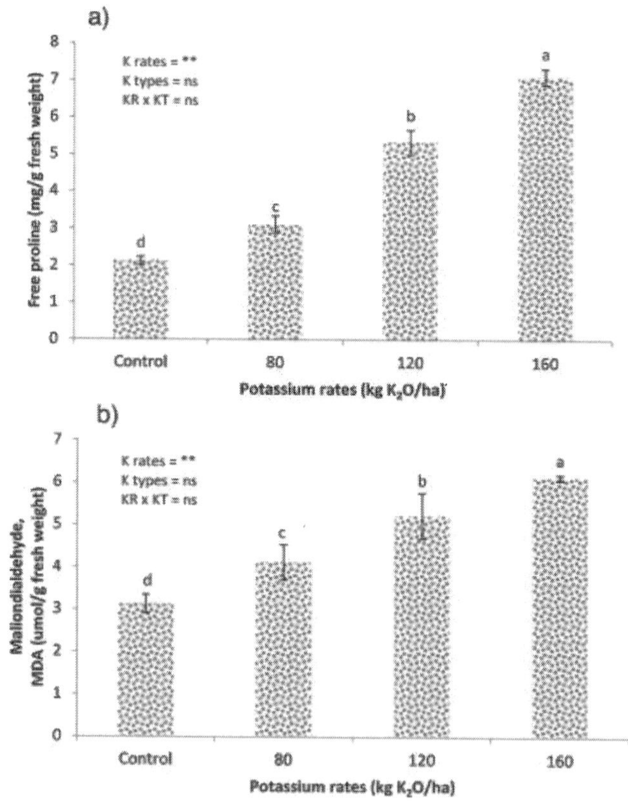

Figure 6. Impact of potassium rates on free proline (a) and Maliondialdehyde content (b) of rice. $N = 28$. Bars represent standard error of differences between means. Control = 80 kg K_2O ha^{-1} KCl + control flooded, ** = significant at 1% level, ns = no significant ($p < 0.05$).

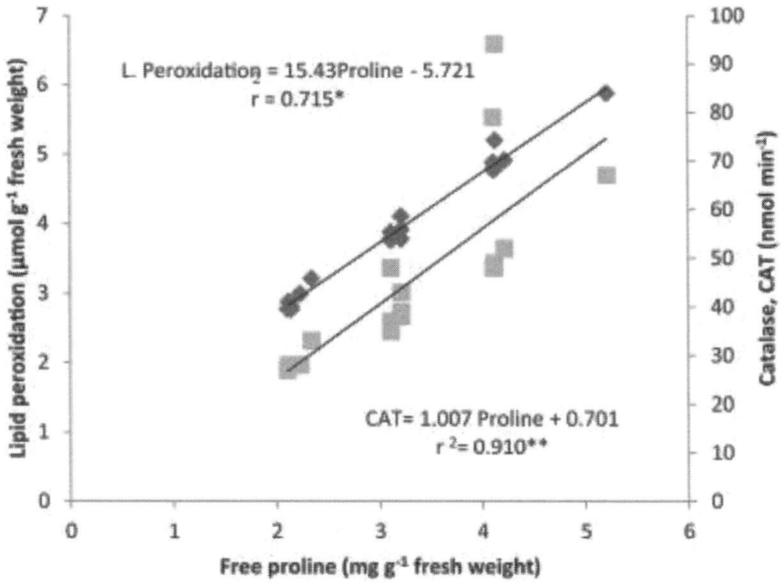

Figure 7. Relationship between free proline content, lipid peroxidation and catalase activity. $N = 28$. * and ** significant at $p \le 0.05$ and $p \le 0.01$ respectively.

3.8. Transpiration rate and water use efficiency

Transpiration rate was enhanced by interaction effects between potassium rate and types ($p \le 0.05$; Fig. 8). Generally, at 80 kg K_2O ha^{-1} potassium, transpiration rate in KCl (8.52 mmol m^{-2} s^{-1}) and K_2SO_4 (9.01 mmol m^{-2} s^{-1}) was the lowest compared to the control that recorded 9.66 mmol m^{-2} s^{-1}. As the potassium rate increased from 80 > 120 > 160 kg K_2O ha^{-1} the transpiration rate was observed to be increased in both KCl and K_2SO_4, with the highest transpiration rate recorded in KCl + 160 kg K_2O ha^{-1}(12.03 mmol m^{-2} s^{-1}). The result suggested that high potassium rates would reduce water stress effects by having high transpiration rate. High transpiration rate would increase the nutrient uptake that would repair the damage tissue under water stress thus reduce the oxidative stress of rice under water stress condition. This been showed by high significant negative correlations of transpiration rate with CAT activity ($r = -0.871; p \le 0.05$; Table 2). The increased in transpiration rate with increased in potassium fertilization have been observed by the other researches (Xu et al., 2011, Kanai et al., 2011 and Sritontip et al., 2008). In 80 and 120 kg K_2O ha^{-1}, it was found that the water use efficiency was

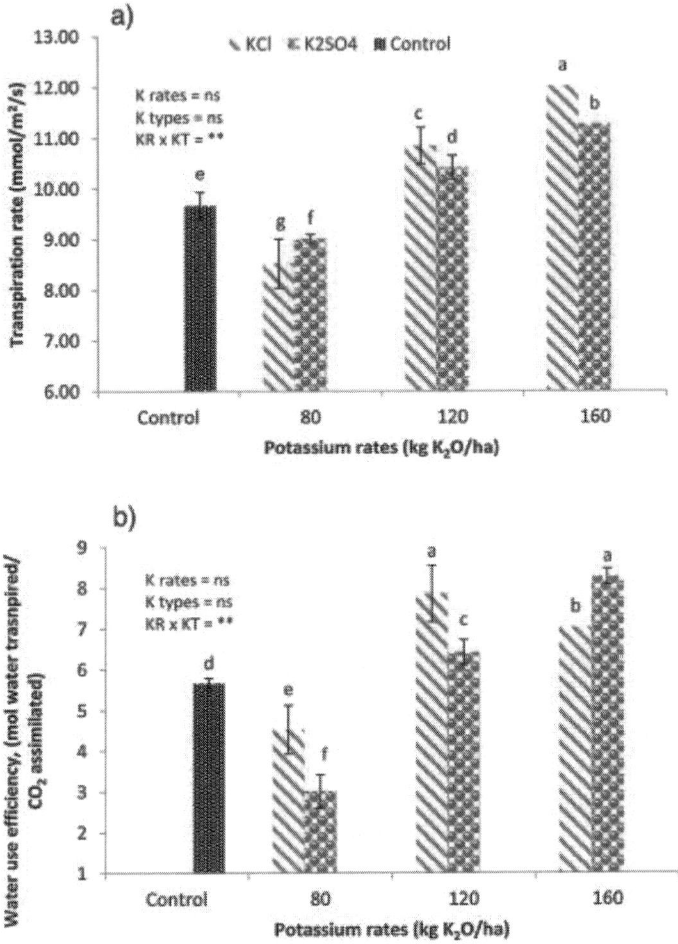

Figure 8. Interaction effects between potassium rates and sources on transpiration rates (a) and water use efficiency (b) of rice. $N = 28$. Bars represent standard error of differences between means. Control = 80 kg K_2O ha^{-1} KCl + control flooded. ** = significant at 1% level, ns = no significant ($p < 0.05$).

highest under application of KCl than K_2SO_4 that recorded 4.52 and 7.85 in 80 and 120 kg K_2O ha^{-1} respectively compared to 3.01 and 6.41 in 80 and 120 kg K_2O ha^{-1} respectively in K_2SO_4. At 160 kg K_2O ha^{-1}, the application of K_2SO_4gave the highest WUE (8.27) compared to KCl that registered 7.03. The control treatment just recorded 5.66. The current study suggests that application of potassium at higher rates can improve the water use efficiency in rice. The increase in water use efficiency in rice might be contributed by efficient turgor regulation and osmotic adjustment under high potassium application in rice (Marshner, 1995). The water use efficiency have a significant correlationship with rice yield ($r = 0.887$; $p \le 0.05$; Table 2) and Harvest index ($r = 0.910$; $p \le 0.05$; Table 2) that indicate the alleviation of drought stress in rice under potassium fertilization might be due to increase in water use efficiency. This was supported by previous studies that showed increased application of potassium would increase the efficiency of plant to use water.

4. CONCLUSION

The application of potassium fertilizers can increase the tolerance of rice under cyclic water stress. The application of potassium at 120 kg K_2O ha^{-1} either KCl or K_2SO_4fertilizer was found to be efficient to minimize reduction in rice yield and harvest index under cyclic water stress 15 days and possible to substitute control flooded practice during water scarcity. It was found potassium might tolerate to cyclic water stress by increase its transpiration rate, net assimilation rate, proline and maliondialdehyde and reduce catalase activity as indication of low oxidative stress under these condition.

ACKNOWLEDGEMENTS

The authors are grateful to the Ministry of Higher Education Malaysia and Long-term Research Grant Scheme (LRGS) for financing this work under the Research University Grant.

REFERENCES

1. Abd-Alla, M.H., Wahab, A.M.A., 1995. Response of nitrogen fixation, nodule activities, and growth to potassium supply in water-stressed broad bean. J. Plant Nutr. 18, 1391–1402.

2. Aebi, H., 1983. Catalase. In: Bergmeyer, H., Verlag, C., Weinheim, A. (Eds.), Methods of Enzymatic Analysis. , 3rd ed. Verlag Chemie, Weinheim, Germany, pp. 273–277.

3. Aref, F., Rad, H.E., 2012. Physiological characterization of rice under salinity stress during vegetative and reproductive stages. Indian J. Sci. Technol. 5, 2578–2586.

4. Asch, F., Dingkuhn, M., Sow, A., Audebert, A., 2005. Drought-induced changes in rooting patterns and assimilate partitioning between root and shoot in upland rice. Field Crops Res. 93, 223–236.

5. Banoc, D.M., Yamauchi, A., Kamoshita, A., Wade, L.J., Pardales, J.R., 2000. Genotypic variations in response of lateral root development to fluctuating soil moisture in rice. Plant Prod. Sci. 3, 335–343.

6. Benlloch-Gonzalez, M., Arquero, O., Fournier, J.M., Barranco, D., Benlloch, M., 2008. K+ starvation inhibits water-stress-induced stomatal closure. Plant Physiol. 165, 623–630.

7. Beringer, H., Trolldenier, G., 1978. Influence of K nutrition on the response to environmental stresses. In: Potassium Research—Reviews and Trends. International Potash Institute, Basel, Switzerland, pp. 189–222.

8. Cakmak, I., 1997. Role of potassium in protecting higher plants against photo-oxidative damage. In: Johnston, A.E. (Ed.), Food Security in the WANA Region, the Essential Need for Balanced Fertilization. International Potash Institute, Basel, Switzerland, pp. 345–352.

9. Chaitanya, K., Sundar, D., Masilamani, S., Ramachandra, R.A., 2002. Variation in heat stress induced antioxidant enzyme activities among three mulberry cultivars. Plant Growth Regul. 36, 175–180.

10. Colmer, T.D., Gibberd, M.R., Wiengweera, A., Tinh, T.K., 1998. The barrier to radial oxygen loss from roots of rice (Oryza sativa L.) is induced by growth in stagnant solution. Exp. Bot. 49, 1431–1436.

11. Egila, J.N., Davies Jr., F.T., Drew, M.C., 2001. Effect of potassium on drought resistance of Hibiscus rosa-sinensis cv. Leprechaun: plant growth, leaf macro and micronutrient content and root longevity. Plant Soil 229, 213–224.

12. Ford, C.W., Wilson, J.R., 1981. Changes in levels of solutes during osmotic adjustment to water stress in leaves of four tropical pasture species. Aust. J. Plant Physiol. 8, 77–91.

13. Foyer, C.H., Decourviers, P., Kenerk, K.J., 1994. Protection against oxygen radicals: an important defense mechanism studied in transgenic plants. Plant Cell Environ. 17, 507–523.

14. Fukai, S., Cooper, M., 1995. Development of drought-resistant cultivarsusing physio-morphological traits in rice. Field Crops Res. 40, 67–86.

15. Hsio, T.C., 1973. Plant responses to water stress. Ann. Rev. Plant Physiol. 24, 519–570.

16. Ibrahim, M.H., Jaafar, H.Z.E., 2012. Primary, secondary metabolites, H2O2, malondialdehyde and photosynthetic responses of Orthosiphon stamineus Benth. to different irradiance levels. Molecules 17, 1159–1176.

17. Itoh, R., Kumara, A., 1987. Acclimation of soybean plants to water deficit. V. Contribution of potassium and sugar to osmotic concentration in leaves. Jpn. J. Crop Sci. 56, 678–684.

18. Kanai, S., Moghaieb, R.E., El-Shemy, H.A., Panigrahi, R., Mohapatra, P.K., Ito, J., Nguyen, N.T., Saneoka, H., Fujita, K., 2011. Potassium deficiency affects water status and photosynthetic rate of the vegetative sink in green house tomato prior to its effects on source activity. Plant Sci. 180, 368–374.

19. Kant, S., Kafkafi, U., 2002. Potassium and Abiotic Stresses in Plants. In: Pasricha, N.S., Bansal, S.K. (Eds.), Potassium for Sustainable Crop Production. Potash Institute of India, Gurgaon, India, pp. 233–251.

20. Khadr, M.S., Negm, A.Y., Khalil, F.A., Antoun, L.W., 2004. Effect of potassium chloride in comparison with potassium sulfate on sugar cane production and some soil chemical properties under Egyptian conditions. Rabat, Morocco, 24–28 November In: IPI Regional Workshop on Potassium and Fertigation development in West Asia and North Africa, 004, pp. 1–8.

21. Khan, A.W., Mann, R.A., Saleem, M., Majeed, A., 2012. Comparative rice yield and economic advantage of foliar KNO3 over soil applied K2SO4. Pak. J. Agric. Sci. 49, 481–484.

22. Lindhauer, M.G., 1985. Influence of K nutrition and drought and water stressed sunflower plants differing in K nutrition. J. Plant Nutr 10, 1965–1973.

23. Liu, S.P., Zheng, L.Q., Xue, Y.H., Zhang, Q.A., Wang, L., Shou, H.X., 2010. Overexpression of OsVP1 and OsNHX1 increases tolerance to drought and salinity in rice. Plant Biol. 53, 444–452.

24. Marschner, H., 1995. Mineral Nutrition of Higher Plants, 2nd ed. Academic Press, San Diego, California, USA.

25. Molden, D., Murray-Rust, H., Sakthivadivel, R., Makin, I., 2003. A water-productivity framework for understanding and action. Water productivity in agriculture: limits and opportunities for improvement. In: Kijne, J.W., Barker, R., Molden, D. (Eds.), Comprehensive Assessment of Water Management in Agriculture Series, No. 1. International Water Management Institute, Sri Lanka.

26. Nandwal, A.S., Hooda, A., Datta, D., 1998. Effect of substrate moisture and potassium on water relations and C, N and K distribution in Vignaradiata. Biol. Plant. 41, 149–153.

27. Pervez, H., Ashraf, M., Makhdum, M.I., 2004. Influence of potassium nutrition on gas exchange characteristics and water relations in cotton (Gossypium hirsutum L.). Photosynthetica 42, 251–255.

28. Pier, P.A., Berkowitz, G.A., 1987. Modulation of water stress effects on photosynthesis by altered leaf K+. Plant Physiol. 85, 65–661.

29. Premachandra, G.S., Saneoka, H., Ogata, S., 1991. Cell membrane stability and leaf water relations as affected by potassium nutrition of water-stressed maize. Exp. Bot. 42, 739–745.

30. Price, A.H., Tomos, A.D., 1997. Genetic dissection of root growth in rice (Oryza sativa L.). II. Mapping quantitative trait loci using molecular markers. Theor. Appl. Gene 95, 143–152.

31. Rad, H.E., Aref, F., Rezaei, M., 2012. Rice growth and yield components respond to changes in water salinity stress. World Appl. Sci. J. 20, 997–1007.

32. Rama Rao, N., 1986. Potassium nutrition of pearl millet subjected to moisture stress., J. Potassium Res. 2, 1–12.

33. Sen Gupta, A., Berkowitz, G.A., Pier, P.A., 1989. Maintenance of photosynthesis at low leaf water potential in wheat. Plant Physiol. 89, 1358–1365.

34. Shahid, U., 2006. Alleviating adverse effects of water stress on yield of sorghum: mustard and groundnut by potassium application. Pak. J. Bot. 38, 1373–1380.

35. Sharma, K.D., Nandwal, A.S., Kuhad, M.S., 1996. Potassium effects on CO2 exchange, NRA and yield of cluster bean cultivars under water stress. J. Potassium Res. 12, 412–423.

36. Siddiqui, M.H., Al-Whaibi, M.H., Sakran, A.M., Basalah, M.O., Ali, H.M., 2012. Effect of calcium and potassium on antioxidant system of Vicia faba L. under cadmium stress. Int. J. Mol. Sci. 13, 6604–6619.

37. Sritontip, C., Khaosumain, Y., Changjeraja, S., Changjeraja, R., 2008. Effects of light intensity and potassium chlorate on photosynthesis and flowering in 'Do' longan. Acta Hortic. 787, 285–288.

38. Su, D., Chen, N., Gao, T., Wang, C., Sheng, M., Yang, C., 2012. Effects of Si+, K+, and Ca2+ on antioxidant enzyme activities and osmolytes in Halocnemum strobilaceum under salt stress. Adv. Mater. Res. 356–360, 2542–2550.

39. Tanner, C.B., Sinchlair, T.R., 1983. Efficient water use in crop production reaseach. In: Taylor, H.M., Jordon, W.R., Sinchlair, T.R. (Eds.), Limitation to Efficient Water Use in Crop Production. ASA/CSAA/ASSA, Wisconsin, pp. 1–8.

40. Tiwari, H.S., Agarval, R.M., Bhatt, R.K., 1998. Photosynthesis, stomatal resistance and related characters as influenced by potassium under normal water supply and water stress condition in rice (Oryza sativa L.). Indian J. Plant Physiol. 3, 314–316.

41. Tripathi, B.N., Bhatt, I., Dietz, K.J., 2009. Photosynthetic organism. Protoplas 235, 3–15.

42. Umar, S., 2006. Alleviating adverse effects of water stress on yield of sorghum, mustard and groundnut by potassium application. Pakistan J. Bot. 38, 1373–1380.

43. Umar, S., Bansal, S.K., 1997. Effect of potassium application on water stressed groundnut. Fert. News 42, 27–29.

44. Umar, S., Moinuddin, 2002. Genotypic differences in yield and quality of groundnut as affected by potassium nutrition under erratic rainfall conditions. J. Plant Nutr 25, 1549–1562.

45. Xu, Y.W., Zou, Y.T., Husaini, A.M., Zeng, J.W., Guan, L.L., Liu, Q., Wu, W., 2011. Optimization of potassium for proper growth and physiological response of Houttuynia cordata Thunb. Environ. Exp. Bot. 71, 292–297.

46. Yadav, D.S., Goyal, A.K., Vats, B.K., 1999. Effect of potassium in Eleusine coracana (L.) Gaertn. Under moisture stress conditions. J. Potas. Res. 15, 131–134.

47. Zeng, Q., Liu, B., Gilna, B., Zhang, Y., Zhu, C., Ma, H., Pang, J., Chen, G., Zhu, J., 2010. Elevated CO2 effects on nutrient competition between a C3 crop (Oryza sativa L.) and a C4 weed (Echinochloacrus-galli L.). Nutr. Cycl. Agroecosyst. 89, 93–104.

8

CHAPTER

On the Conservative Behavior of Biomass Water Productivity

Pasquale Steduto , Theodore C. Hsiao ,Elı`as Fereres

[1]Division of Land and Water, FAO,United Nations, viale delle Terme di Caracalla,00153 Rome, Italy

[2]Department of Land, Air and Water Resources, University of California, Davis, CA, USA

[3]IAS-CSIC and University of Cordoba, Cordoba, Spain

ABSTRACT

The ever-increasing demand and competition for the finite water resource worldwide call for more efficient use of water in all sectors, including firstly agricultural food production. One important consideration is the existence of a limit to the amount of biomass a crop can produce per unit of water consumed. This article analyzes the theoretical background and the experimental evidence for the conservative behavior of the efficiency in water use by crops to produce biomass, i.e., biomass water productivity (WP_b), under variable environmental conditions. Particularly, WP_b is approximately constant for a given crop species after normalization for evaporative demand of the atmosphere and air carbon dioxide concentration. A stepwise scaling up approach, from leaf to canopy, is undertaken to underline the processes involved at the different hierarchical levels of biological organization that lead to the conservative behavior of WP_b. Starting at the leaf level, the basic gas exchange equations are outlined to demonstrate that the normalized photosynthetic WP_b at the leaf scale is proportional to the ambient CO_2 concentration. New experimental evidence in support of that conclusion is presented for several C_3 and a C_4 crops. Additional factors are introduced to assess photosynthetic WP_b at the canopy scale, including the extent of radiation capture and the role of respiration. The composition of biomass was then considered in the analysis of WP_b over a season. The paper highlights the need to normalize

WP_b for differences in climate, specifically, in evaporative demand of the atmosphere to extrapolate WP_b values between climatic zones, and in atmospheric CO_2 concentration to account for changes in CO_2 with time, when looking at the past and into the future. Two procedures for normalization for differences in evaporative demand are presented, and a procedure for normalization for changes in CO_2 concentration is derived for the leaf scale and shown to be applicable to canopy scale. Some knowledge gaps and research needs are pointed out and the potential offered by the near constancy of normalized WP_b in crop simulation modeling is emphasized.

INTRODUCTION AND BACKGROUND

Food production and water use are two closely linked processes. As the competition for water intensifies worldwide, water in food production must be used more efficiently. Of the different steps in water use in the crop production process, the most fundamental is the exchange of water lost by transpiration for the assimilation of carbon dioxide. The net gain of carbon and energy by the plant in this process then leads to the production of biomass, of which the harvested yield is usually only a part. It turns out that for biomass production, the efficiency of water use (transpiration) is relatively constant after the variation in two key environmental factors, evaporative demand of the atmosphere and air carbon dioxide concentration, are accounted for by normalization.

This conservative behavior has been analyzed and discussed in detail several times in the past half century. In light of the urgent need to answer the question of just how much the efficiency of water use in agriculture can be improved, we are revisiting the issue here, to examine the theoretical basis for the conservative behavior, to evaluate just how conservative biomass water productivity is, and to consider the possibility of its improvement. The conceptual basis for the conservative behavior will be reviewed, recent data evaluated and discussed, and the different ways to normalize for evaporative demand and carbon dioxide concentration illustrated. It is hoped that this discourse will help to focus better the potential means to improve the efficiency of water use, and also lead to the modeling of crop productivity based on water use. The subject of this paper, biomass water productivity, is a segment in the sequence of steps in the use of water for crop production. The broader aspects of the efficiencies in agriculture include many other efficiency steps, and these are considered in another paper in this special issue (Hsiao et al. 2007).

Biological or primary productivity is normally evaluated in terms of biomass (as dry matter), and the water consumed and not recoverable in this production process is normally assessed in terms of evapotranspiration (ET), the sum of transpiration by the crop (T) and evaporation from the soil (E). This paper focuses on the production of biomass in relation to transpiration. Only the water transpired is considered because evaporation from the soil is not in exchange for carbon assimilated. Here biomass water productivity (WP_b) is defined as the aboveground dry matter (g or kg) produced per unit land area (m ^{2} or ha) per unit of water transpired (mm or m^3). The units of WP_b are then either kg m^{-3} or kg ha^{-1} mm^{-1}. In the literature water productivity is mostly expressed as the ratio of biomass to evapotranspired water, because separation of E from T was not possible in these cases. Although the units remain the same regardless whether T or ET is the denominator, it is important to specify which water productivity is being referred to.

The scales considered here range from individual leaves up to crop plant communities, with the boundary conditions for the latter being that of an agricultural field. Only above-ground biomass is considered, even though root biomass is a reasonably significant component of total biomass and a fraction of daily photosynthate is allocated to the growth and maintenance of the root system (Lambers 1987). The very limited information on root biomass in relation to water use, relative to that existing on shoot biomass, made it necessary to restrict this analysis to shoot biomass production. Fortunately, for most crop species except root crops, only a small portion of the total biomass is in roots so that considerable changes in root biomass may occur with only minor effect on total biomass. Relative to shoot, root growth is known to be more resistant to nutrient and water deficiencies (Taylor 1983; Hsiao and Xu 2000a). On the other hand, there is a homeostatic growth response toward a near constant root:shoot ratio (Brouwer 1983).

Water productivity is also referred to as water use efficiency (WUE) in the literature. Monteith (1984, 1993) criticized the WUE term and pointed out that no theoretical limits exist as reference, as it should be for efficiency in an engineering sense. On the other hand, efficiency is used widely in economics without a theoretical maximum. More practically, the extensive use of WUE had caused confusion and met objections mainly because the meaning of the term depends on the specifics of the numerator and denominator defining it. This problem can be overcome by specifying with subscripts, as exemplified in Hsiao et al. (2007).

Many experiments have shown that the relationship between biomass produced and water consumed by a given species is highly linear. This indicates that WP_b is approximately constant, a feature that is critical for the analyses of water-limited productivity. The linearity between crop biomass (and often final yield) and water use has been observed since the early 1900 (e.g., Briggs and Shantz 1913a, b), although the first analytical approach to formalize the relationship was developed by de Wit (1958). Several hundreds of linear relationships can be found in the literature (e.g., Hanks 1983) along with different approaches to link both variables (e.g., Arkley 1963; Bierhuizen and Slatyer 1965; Stewart 1972; Hanks 1974; Stanhill 1986). Tanner and Sinclair (1983) presented a systematic analysis that provided a theoretical basis and confirmed the previously observed constancy of WP_b for a given environment. Since then, only few attempts have been made to combine the scientific advances and the new experimental evidence to improve our understanding of the behavior of WP_b [e.g., Hsiao (1993) for high CO_2; Steduto (1996) and Steduto and Albrizio (2005) for climate normalization procedures]. This is surprising in view of the practical implications that a constant WP_bwould have for the use of a limited amount of water in supplemental irrigation (Oweis et al. 2000) and in regulated deficit irrigation (Fereres et al. 2003), as well as in the theoretical approaches to modeling crop production (Steduto 1996).

Actually, the realm of dynamic crop-growth modeling has evolved based on the use of another fundamental and conservative parameter—radiation use efficiency (RUE)—the slope of the relationship between biomass produced and solar radiation intercepted (or absorbed) by the crop canopy. The choice of the RUE formalism (Monteith 1977) was likely influenced by the relative ease by which RUE can be determined as compared to WP_b, and by the fact that the first modeling effort focused on potential yield (de Wit et al. 1970), which is radiation-limited rather than water-limited. Nevertheless, the ever increasing recognition of water as a limiting resource worldwide (Seckler et al. 1998), the knowledge accumulated on crop-water relations and water productivity, and the recent improvements in methods for the determination of WP_b all argue for more attention be given to the potential offered by WP_b in crop simulation models. This is supported by the variability encountered in RUE (Sinclair and Muchow 1999; Albrizio and Steduto 2005) along with the better simulation performance in using WP_b as compared to using RUE observed in a few cases (Steduto and Albrizio 2005).

THEORETICAL FRAMEWORK AND EXPERIMENTAL EVIDENCE

In developing the theoretical background and the appropriate framework for analyzing the constancy of WP_b, we follow a stepwise scaling-up approach, from leaf to crop community, in the analysis of the two basic processes involved, water transpiration (*T*) and net carbon assimilation (*A*), and its conversion to biomass.

Photosynthetic water productivity—leaf scale

At the leaf level, we define photosynthetic water productivity (WP_p) as the ratio of leaf net carbon dioxide assimilation (A_l) to leaf transpiration (T_l), both expressed as flux rates on a leaf area basis (mol m^{-2} s^{-1}). In the gas exchange processes between a leaf and its environment, CO_2 and water vapor share the same pathway between the bulk atmosphere and the intercellular air space. While this completes the path for water vapor, CO_2 has yet to move in liquid phase from the cell walls to the carboxylating sites of the thylakoids (C_3species) or cytosol (C_4 species). This additional path for CO_2 may be ignored when gas exchange is determined under steady state or near steady state conditions. Neglecting the cuticular path and assuming steady state, the corresponding gaseous A_l and T_l fluxes are expressed as:

$$A_l = \frac{\Delta c}{r'_b + r'_s} = \frac{c_a - c}{r'_b + r'_s} \quad (1)$$

$$T_l = \frac{\Delta w}{r_b + r_s} = \frac{w_i - w}{r_b + r_s} \quad (2)$$

where Δc is the difference in CO_2 concentration between that of the atmosphere (c_a) and that in the leaf intercellular air space (c_i); Δw is the water vapor concentration difference between the leaf intercellular air space (w_i) and the atmosphere (w_a); r'_b and r'_s are the boundary layer and stomatal resistances, respectively, for CO_2 transport; and r_b and r_s are the boundary layer and stomatal resistances, respectively, for water vapor transport.

Equations 1 and 2 express the A and T fluxes in purely physical terms while all the complex metabolic processes of CO_2 fixation at the biochemical level are imbedded in the c_i term of Δc.

Along the path described, both gases encounter the same resistance to diffusion except for their different binary diffusivity due to a difference in molecular mass (von Caemmerer and Farquhar 1981). The relationship between the total resistance to water vapor transport ($r = r_b + r_s$) and the total resistance to CO_2 transport ($r' = r'_b + r'_s$) is about constant under high turbulence (a condition typical of gas-exchange measurements in cuvettes, and very common in open fields). Farquhar and Sharkey (1982) determined that $r \cong 0.625r'$. This constant relationship indicates that any change of the resistance to gas transport will have a similar impact on A_l and T_l.

Under steady-state conditions then, photosynthetic water productivity of single leaves is expressed as:

$$\mathrm{WP_p} = \frac{A_l}{T_l} = \frac{(r_b + r_s)}{(r'_b + r'_s)} \frac{\Delta c}{\Delta w} = \frac{r}{r'} \frac{\Delta c}{\Delta w} \cong 0.625 \frac{\Delta c}{\Delta w} \quad (3)$$

Equation 3 underlines the dependence of $\mathrm{WP_p}$ on the concentration difference or gradient of the two gases. At any given Δw, an increase in Δc under elevated atmospheric CO_2 will increase $\mathrm{WP_p}$. Similarly, $\mathrm{WP_p}$ is influenced by the evaporative demand of the atmosphere through Δw and will vary in different climates unless it is normalized for Δw. At the leaf scale, this normalization is quite simple, as Δw is determined during the flux measurements of gas-exchange.

To normalize $\mathrm{WP_p}$ for Δw, both sides of Eq. 3 are multiplied by Δw. That will cancel the denominator of Eq. 3and make the normalized $\mathrm{WP_p}$ dependent only on Δc. Also, if c_a is maintained constant by the gas-exchange system, the normalized $\mathrm{WP_p}$ ($\mathrm{WP_p}^*$) will only be influenced by the degree of variation in c_i.

Normalizing for Δw and rearranging,

$$\begin{aligned} \mathrm{WP_p^*} &= 0.625 \frac{\Delta c}{\Delta w} \Delta w = 0.625\,(c_a - c_i) \\ &= 0.625\,c_a - 0.625\,c_i = k - 0.625\,c_i \end{aligned} \quad (4)$$

where $k = 0.625\,c_a$. Equation 4 shows that for a given c_a, $\mathrm{WP_p}^*$ is a linear function of c_i with an intercept of k ($= 0.625\,c_a$) and a slope of 0.625. That is, under a given c_a and evaporative demand, $\mathrm{WP_p}$ is critically dependent on c_i.

As depicted in Fig. 1, c_i represents a balance between the importation of CO_2 into the intercellular space through stomata and air boundary layer, and the depletion or utilization of the intercellular CO_2 by photosynthetic carboxylation. The rate of importation is jointly determined by the level of c_a relative to c_i(Δc) and the relevant resistances (r'_b and r'_s) in the gaseous transport pathway. The rate of depletion is jointly determined by the level of c_i, the activities of the enzymes controlling the CO_2 dissolution and fixation processes in photosynthesis, and the level of the substrates as CO_2 acceptors. The lumped resistance r'_m in the figure represents essentially the overall biochemical resistance to CO_2 fixation. Plants have apparently evolved feedback and feedforward mechanisms (dashed lines) to keep the importation and depletion of CO_2in balance most of the time so that c_i is conservative. This implies that photosynthetic capacity and stomatal opening are coordinated and operate in concert in the leaf. Under fast changing conditions, however, c_i does fluctuate but then gradually returns close to its set value. This suggests that when one of the two opposing processes, either the importation or depletion of CO_2, is perturbed, the other adjusts with some lag to keep the system in balance and c_i nearly constant.

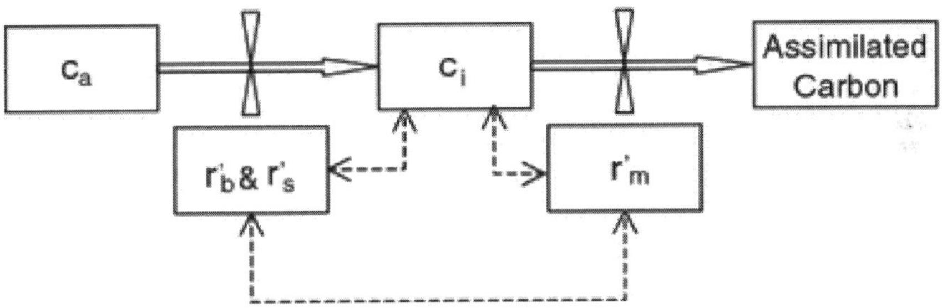

Figure 1 Schematic view of leaf intercellular CO_2 concentration (c_i) as a balance between the importation and depletion of CO_2. Importation of ambient CO_2 (c_a) is regulated by boundary layer (r'_b) and stomatal (r'_s) resistances, and depletion of CO_2 by carbon assimilation is dependent on the biochemistry of photosynthesis (r'_m). Control loops (feedback and feed forward), indicated by the *dashed lines,* appear to have evolved to optimize the gas exchange processes, resulting in a conservative c_i

There has been substantial experimental evidence showing that for many species, c_i tends to remain constant under a range of conditions (Wong et al. 1979; Pearcy 1983; review by Morrison 1987; Hsiao and Jackson 1999) including water stress when the stress develops gradually, as it generally occurs in the field. It is also well known that there is a difference in c_i between the leaves of C_3 and C_4 species, due to differences in their photosynthetic pathways (Osmond et al. 1982). Near constancy of c_i has been observed with variations in temperature (e.g., Björkman 1981), radiation (e.g., Bolaños and Hsiao 1991), water supply (e.g., Wong et al. 1979; Hirasawa et al. 1995), leaf nitrogen content (e.g., Wong et al. 1979) and salinity stress (Steduto et al. 2000). For species with stomatal response to humidity, however, as water vapor gradient from the leaf interior to the air (Δw) increases stomata react by narrowing and c_i decreases (Morison 1987). Such responses were apparently influenced by changes in leaf temperature and plant water status (e.g., Steduto et al. 1997) and will be discussed further in "Some unresolved issues".

The ample evidence under most circumstances on the tendency of c_i to remain constant at a constant c_a, i.e., a constant c_i/c_a ratio, is an indication that stomata behave at the leaf scale in a manner that leads to a constant WP_p^*.

By considering the ratio $c_i/c_a = \alpha$ as constant, $c_i = \alpha c_a$, and Eq. 4 can be arranged as follows

$$WP_p^* = 0.625\,c_a - 0.625\,c_i = 0.625\,(1-\alpha)\,c_a = k^* c_a \tag{5}$$

clearly indicating that WP_p^* is proportional to the increase in ambient CO_2 by the proportionality of k^*, as long as α stays constant. Using the widely accepted generalized values for α of 0.7 and 0.4 for C_3 and C_4species, respectively (Morison 1987; Wong et al. 1979), k^* and WP_p^* take on the following values:

$$k^* = 0.1875 \quad WP_p^* = 0.1875\,c_a \quad \text{for } C_3$$
$$\text{and} \quad k^* = 0.375 \quad WP_p^* = 0.375\,c_a \quad \text{for } C_4.$$

k^* is unitless and the units for WP_p^* are the same as that for c_a and c_i.

Figure 2 presents a comparison of three sets of WP_p^* values spanning a range of c_a for several field grown crops. One set was experimentally determined; one set was calculated with Eq. 4 using the experimentally measured c_i that corresponded to the experimental c_a values (plotted), and another was calculated from $c_a$using Eq. 5. The general agreement found among the three sets of WP_p^* for these crops, particularly over the range of c_a that is not excessively high, confirms the validity of the assumptions made to arrive at the simple expressions defining normalized photosynthetic water productivity at the leaf scale (WP_p^*). Most notable is the fact that Eq. 5 makes no differentiation among crop species except between C_3 and C_4 and there is no parameter to adjust regardless of the crop, and no experimental measurement to make as long as c_a is known. Yet the equation yielded WP_p^* matching closely the measured values for the C_3 species wheat, sunflower, sugar beet and olive, and the C_4 species sorghum, over a wide range of c_a. It should be mentioned though, that the measured and the simulated responses tend to diverge at high c_a. This is most likely because the capacity of leaves to maintain c_i constant diminishes as CO_2 saturation is approached, as may be expected when c_a was changed quickly in these short-term experiments, with little opportunity for acclimation by the leaves.

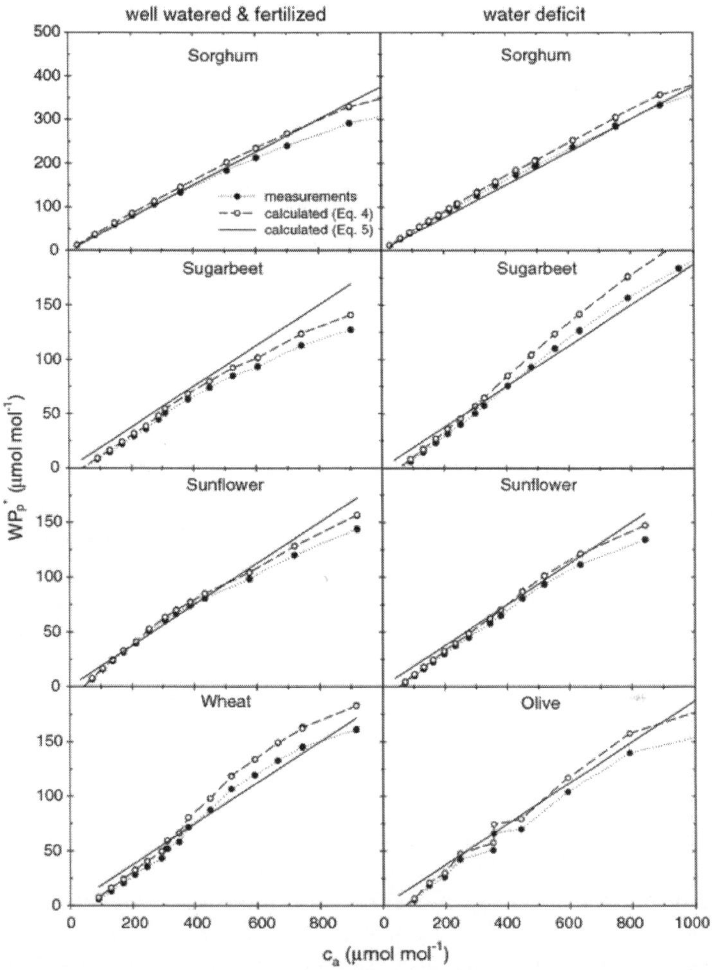

Figure 2. Examples of leaf photosynthetic water productivity ($WP_p{}^*$), normalized for Δw, as dependent on ambient CO_2 concentration (c_a), for different crops under well watered and high nitrogen (*left plots*) and water deficient and low nitrogen (*right plots*) conditions. Measurements (*dotted line with filled circles*) were obtained from the determination under steady state conditions of A_l versus CO_2 response curves, using a portable leaf-photosynthesis open-system (Li-6400, LiCor, Lincoln, NE, USA) following the procedure described in Steduto et al. (2000). *Dashed line* (with *open circles*) and *continuous lines* (with no symbol) represents values calculated according to Eqs. 4 and 5, respectively. Data from P. Steduto and R. Albrizio (unpublished)

Once again, it bears emphasis that only c_a needs to be known in order to use Eq. 5, and no other measurements are required.

The emphasis here is on the constancy of the ratio of c_i to c_a (α) and the constancy of c_i at a given c_a. Nonetheless, following the discovery of the variation in carbon isotope discrimination and its relation to WP (Farquhar and Richards 1984; Farquhar et al. 1989), efforts have been made to breed C_3 crops for higher WP by selecting for reduced carbon isotope discrimination, which corresponds to a lower α. Obviously genetic variation in α exists. The key question is how much can we expect α to vary in the coming few decades for widely used crop cultivars. So far the variations appear to be small for most of the crop species examined, as discussed further in "Some unresolved issues".

Photosynthetic water productivity—canopy scale

In an agricultural system, assimilation and transpiration by the crop canopy determine the fluxes of CO_2 and water vapor and thus, canopy photosynthetic water productivity (WP_p^c), and ultimately, biomass water productivity (WP_b). As we scale up from leaf to canopy, there are additional features that must be taken into account because the consideration is now on a land area basis instead of leaf area basis. When the coverage of the land by crop canopy is incomplete, the capture of radiation becomes critical for both photosynthesis and water loss.

The commonality and differences in the factors that affect canopy photosynthesis and canopy transpiration are summarized in Fig. 3. The extent of radiation capture by a crop depends on the amount of leaf area, normally evaluated by the leaf area index (LAI), on the geometric arrangement of the leaves within the canopy, as well as on the angle and intensity of incident radiation. Therefore, in addition to solar radiation, plant factors, particularly plant density and the stage of vegetative growth, are critical determinants of radiation interception when the canopy is incomplete. As in leaves, the process of canopy transpiration (T_c) shares the same source of captured energy as the canopy assimilation (A_c). Of the total captured solar radiation though, only the fraction that is photosynthetically active (PAR) is effective in CO_2 assimilation, while the whole spectrum is used for transpiration. PAR, however, is a fairly constant fraction of the incident solar radiation (Meek et al. 1984; Varlet-Grancher et al. 1989) as is the ratio of absorptance of PAR to non-PAR radiation for the leaves of many species (Stanhill 1981). Another fact to recognize is that at the canopy level, A_c of many crops usually does not reach light saturation but has an essentially linear response to irradiance (Biscoe et al. 1975; Louwerse 1980; Asseng and Hsiao 2000). Consequently, any change in the amount of radiation captured by the canopy (e.g., difference in effective leaf surface area, weather, etc.) would affect in a similar way A_c and T_c. Thus, at the canopy level, not only the CO_2 and water vapor share much of the transport pathway, they also share the dominant energy source. There is, however, a difference in that sensible heat flux can either add or remove energy for transpiration from the canopy independent of radiation. Overall though, the sharing of radiative energy source is a critical and often even dominant factor in linking assimilation and transpiration rates at the canopy level.

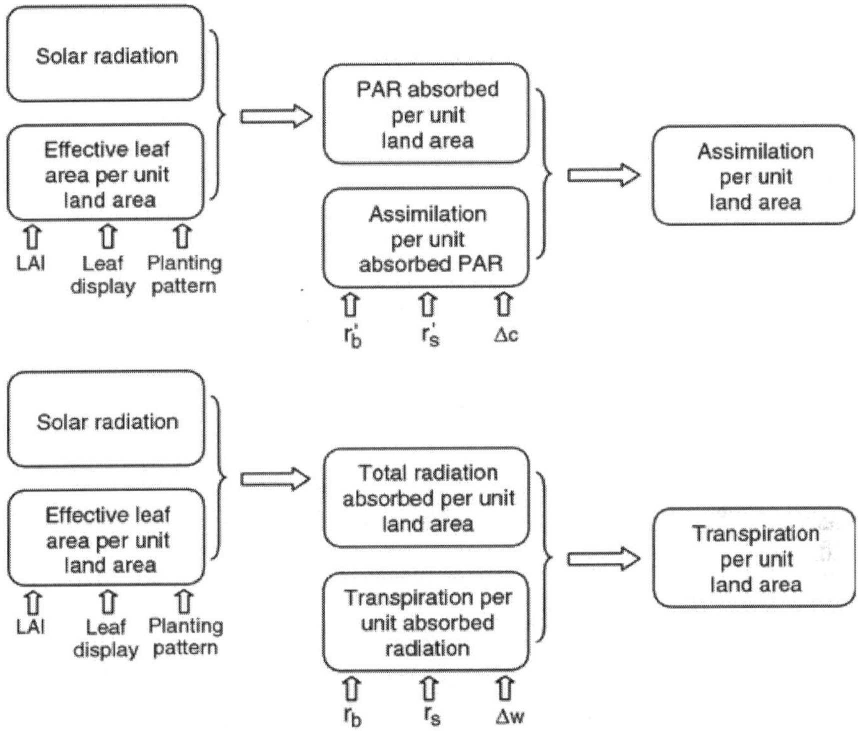

Figure 3. Similarities and differences in factors affecting assimilation and transpiration of canopies. *Arrows* indicate causal relations. All considerations are on a basis of land area. Symbols are: r'_b, boundary layer resistance to CO_2; r'_s, stomatal resistance to CO_2; r_b, boundary layer resistance to water vapor; r_s, stomatal resistance to water vapor; Δc, difference in CO_2 concentration between the atmosphere and the leaf intercellular space; and Δw, difference in water vapor concentration between leaf intercellular space and the atmosphere. Note that because resistances to CO_2 and to water vapor are proportional to each other, most of the factors affecting assimilation have analogous impact on transpiration. One clear difference is the driving force for gas transport, with Δc for assimilation and Δw for transpiration. See text for possible differences caused by PAR being only a portion of the total radiation, and sensible heat flux providing another way to exchange energy. Modified from Hsiao and Bradford (1983) and Hsiao (1993)

Figure 3 also emphasizes the most obvious difference between the assimilation and transpiration processes, namely, the difference in driving force for gas transport, being Δc for CO_2, and Δw for water vapor, two virtually independent variables. This provides some means to improve $WP_p{}^c$, through the manipulation of either or both of these parameters. Equally important, for $WP_p{}^c$ to be conservative in different environments, it must be normalized for Δc and Δw.

Data on canopy assimilation and transpiration, measured simultaneously to determine $WP_p{}^c$ over many days, are not extensive. Figure 4 presents some results for sorghum, wheat, and chickpea obtained in the field. The constancy of $WP_p{}^c$ is obvious from the straight line relationship between ΣA_c and ΣT_c. It is necessary to point out, however, that the ratio A_c/T_c actually varies substantially over diurnal cycles (Xu and Hsiao 2004), mostly because evaporative demand of the atmosphere varies diurnally. These variations apparently average out when data were accumulated over periods of many days, as is the case for the data in Fig. 4.

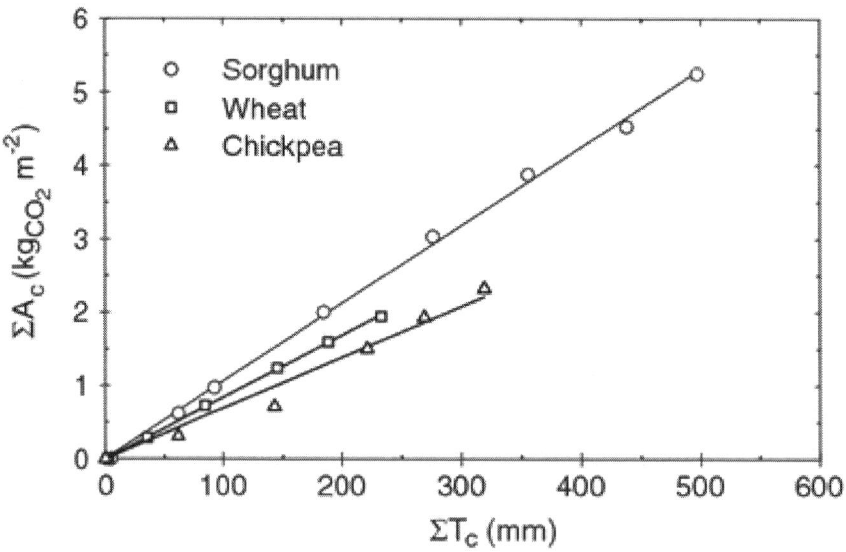

Figure 4. Relationships between cumulative daytime canopy net assimilation (A_c) and cumulative daytime canopy transpiration (T_c) for sorghum, wheat and chickpea. The slope of the relationships represents $WP_p{}^c$. Measurements were taken when the crops were all at full canopy cover, and soil evaporation was assumed to be negligible (redrawn from Steduto and Albrizio 2005)

The data in Fig. 4 are for day time only. As the temporal scale increases from instantaneous measurements of A_c to periods up to a season, more factors must be taken into account. Leaves are not the only components of the canopy. Structural and reproductive organs grow and develop during the crop cycle, changing dimensions, composition, and respiratory requirements as the season progresses. Almost all of the seasonal data in the literature are for biomass water productivity (WP_b). Data on seasonal photosynthetic water productivity ($WP_p{}^c$) are rare because continuous measurement of A_c and $WP_p{}^c$ over a season is extremely difficult to obtain. Changing from A_c to biomass requires an analysis of the respiratory costs in relation to A_c, and of the chemical composition and carbon requirements of the biomass.

Converting the assimilated CO_2 into the products used for growth, maintenance, transport and assimilation of mineral nutrients requires energy supplied by respiration. It is customary to separate respiration into two components, that used in growth and that used in the other processes (so called maintenance). The total respiration of a canopy (R_c) could consume from one fourth to two thirds of the total assimilates over a plant's life cycle (Amthor 1989) and may affect A_c variably over the season. As A_c is the measured net assimilation, variation in R_c with crop ontogeny may affect measure of A_c over the season. The other important consideration is the composition of the biomass. Not only does the biomass composition determine the respiratory growth requirements (Penning de Vries et al. 1974) but also the maintenance costs (Amthor 1989).

All this implies that overall, a linear relationship between A_c and T_c, i.e., a constant $WP_p{}^c$ (e.g., as in Fig. 4) would be expected only if the relationship between R_c and A_c is also linear (Charles-Edwards 1982). Fortunately, more and more evidence is appearing indicating an approximate fixed ratio of assimilation to respiration for a given species or genotype (Amthor 1995). This seems to be the case even as environmental conditions vary, including that in temperature (Gifford 1995), CO_2 concentration (Cheng et al. 2000), and nitrogen availability within a reasonable range (Garcia et al. 1988).

The tight correlations between cumulative A_c and R_c along the season for sunflower, sorghum, wheat, and chickpea are illustrated in Fig. 5. The change in the slope of the linear relationship for sunflower after anthesis in Fig. 5a is the consequence of a significant increase in respiratory requirements due to oil production in the seeds (Whitfield et al. 1989). The increase in CO_2 release through respiration then leads to a lower net canopy assimilation and lower ratio of A_c to R_c. The fact that two different levels of nitrogen yielded the same relationship between A_c and R_c for a C_3 as well as a C_4 species is illustrated by the data for pre-anthesis sunflower (Fig. 5a) and for sorghum (Fig. 5b). These results support the concept of fairly constant carbon use efficiency throughout the crop life cycle, provided that the composition of biomass does not change significantly (as in Fig. 5a). Research on woody plants and forest trees has led to the use of a constant ratio between gross and net primary production in the accurate simulation of the growth of tree stands of many species (Landsberg et al. 2003).

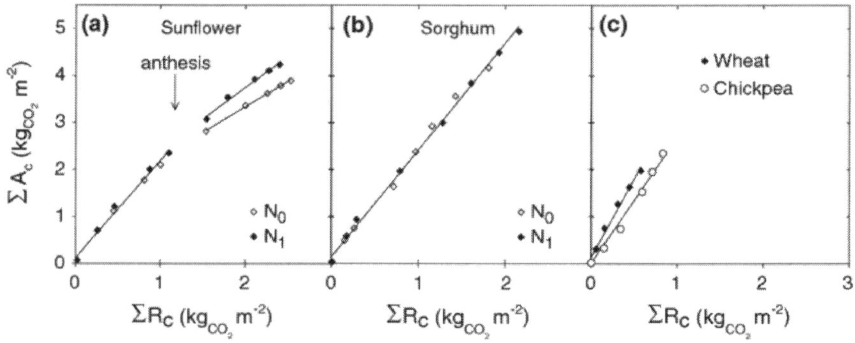

Figure 5. Relationship between cumulative daytime canopy net assimilation (A_c) and cumulative nighttime canopy dark respiration (R_c) for **a** sunflower, **b** sorghum and **c** wheat and chickpea. Slopes of the post anthesis regression lines for N_0 and N_1 are not significantly different at 5% probability (from Albrizio and Steduto2003)

The data in Fig. 4 demonstrated clearly the constancy of photosynthetic water productivity at the canopy scale ($WP_p{}^c$). This is supported and consistent with the underlying factors discussed so far: (a) the conservative behavior of leaf WP_p; (b) the role of solar radiation intercepted by the canopy in determining both A_c and T_c; and (c) the conservative behavior of canopy carbon use efficiency (Fig. 5) provided that chemical composition of the biomass does not change.

From net carbon gain to biomass

The next step is to scale up from daily net A_c to seasonal biomass, in which the only new process influencing biomass accumulation would be the metabolic costs for the construction and maintenance of the biomass. Given that the composition of vegetative parts of many crop species is very similar (Penning de Vries et al. 1983) and does not change substantially, biomass should also be linearly related to transpiration, but the slope of biomass versus transpiration plot (WP_b) will be lower compared to the corresponding plot for $WP_p{}^c$. This is clearly evident with data obtained in the same set of experiments and depicted in Fig. 6. Again, in the case of sunflower (Fig. 6a) the slope during the reproductive phase is less than that in the vegetative phase. Similar behavior should be expected in those crops where the reproductive organ has high protein and/or oil content, such as soybean, peanut (Angus et al. 1983), sunflower, etc. Penning de Vries et al. (1983) provide an assessment of the metabolic costs of production of the harvestable organs for most crops from where it is possible to infer the trend in WP_b depending on the nature of the yield product.

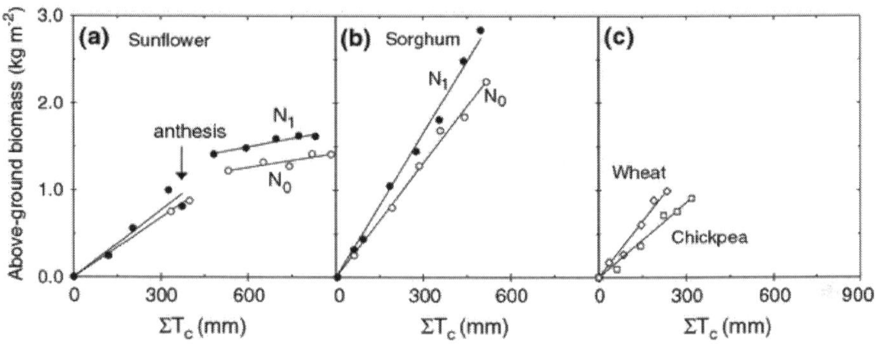

Figure 6. Relationship between cumulative biomass and cumulative canopy transpiration (T_c) for **a** sunflower, **b** sorghum and **c** wheat and chickpea (from Steduto and Albrizio 2005). Sunflower and sorghum data were obtained under two levels of nitrogen nutrition. The differences in the slope of the plot between the two levels of nitrogen is statistically significant only for sorghum

Although the analyses and data above have addressed situations where only above-ground biomass is considered, constant WP_b has been described for root and tuber crops such as sugar beet (e.g., Clover et al.2001), and potato (e.g., Tanner 1981). Presumably, the constancy in water productivity of both above-ground biomass and root biomass in the case of root crops is a consequence of a homeostatic growth response of root and shoot (Brouwer 1983), as earlier indicated. There is, however, a shift under water or nutrient deficiencies in the ratio in favor of the root. But, as already mentioned, this shift has relatively minor impact on crop WP_b because shoot biomass dominates in cases of crops harvested for shoots, and root biomass dominates in cases of crops grown for roots.

Normalization of biomass water productivity for climate

The virtually constant WP_p^c and WP_b^c in Figs. 4 and 6, respectively, were for the cases where environmental conditions did not vary markedly over periods of weeks to months. By analogy with the leaf WP_p, however, one expects WP_p^c and WP_b^c to still be dependent on the magnitudes of the driving forces for water vapor and CO_2 transport, which are determined by the environment. Thus, there is a need to normalize WP_p and WP_b for the climate, specifically, for evaporative demand of the atmosphere to extrapolate water productivity values between climatic zones, and for atmospheric CO_2 concentration to extrapolate between atmospheric CO_2 status. The latter is necessary to evaluate and make use of old data (e.g., those analyzed by de Wit 1958) and to accommodate future rise in atmospheric CO_2.

Normalization for atmospheric evaporative demand

The conceptual normalization of WP_b for the evaporative demand of the atmosphere can take two routes (Tanner and Sinclair 1983; Steduto and Albrizio 2005): (1) normalizing via the "transpiration gradient", or (2) normalizing via "reference transpiration flux".

The gradient-normalizing route is derived from the leaf-scale gas transport model (Eqs. 1, 2). For the canopy, the flux-gradient theory (Norman 1979) is applied considering the canopy as a "big leaf" (Lhomme 1991), with the latent heat flux (λE) expressed as

$$\lambda E = \frac{\rho_a c_p g_w}{\gamma} \Delta e \quad (6)$$

where λ is the specific latent heat of vaporization (J kg^{-1}); E is the evaporation or water vapor flux (kg m^{-2} s^{-1}), all due to transpiration; ρ_a is the air density (kg m^{-3}); c_p is the specific heat capacity of the air (J kg^{-1} °C); g_w is the total conductance for water vapor transport (m s^{-1}), including both aerodynamic and canopy conductances; γ is the psychrometric constant (kPa °C^{-1}); and Δe is the leaf-to-air vapor pressure difference (kPa).

Following the derivation of Tanner and Sinclair (1983) at the canopy level, WP_p^c and WP_b can be normalized for the leaf-to-air water vapor pressure difference (Δe), representing the driving force of the transpiration process (analogous to Δw in Eq. 2), through the following equations

$$\mathrm{WP}_p^{c,*} = \frac{\sum_{i=1}^{n} A_{c,i}}{\sum_{i=1}^{n} t_i \left(\frac{T}{\Delta e}\right)_i} \quad (7)$$

$$\mathrm{WP}_b^{*} = \frac{\text{Biomass}}{\sum_{i=1}^{n} t_i \left(\frac{T_c}{\Delta e}\right)_i} \quad (8)$$

The summations are over a total of *n* number of time intervals of the same length with *i* as the running number designating each interval. The length of the interval (in days) is represented by *t*. Biomass denotes the gain in biomass from the beginning to the end of the summation period. In contrast, A_c, T_c and Δedenote the quantity cumulated during each interval *i*. Ideally, Eqs. 7 and 8 should be applied to the daily data ($t_i = 1$ day) to normalize. Not many studies provide sequential daily data, however, as the majority of the studies employed time intervals of at least 1 week ($t_i =$ 7 days) or longer in the measurements of T_c. Also, with respect to WP_p, t_i often varies from interval to interval (e.g., $t_i = 7$ days, $t_2 = 5$ days, $t_3 = 10$ days, etc.). For practical purposes, Eq. 8 is usually used in an alternative form, namely:

$$\mathrm{WP}_b^* = \frac{\mathrm{Biomass}}{\sum_{i=1}^{n}\left(\frac{T_c}{\overline{\Delta e}}\right)_i} \quad (9)$$

where $\overline{\Delta e}$ ¯is the mean daily Δe for time interval *i*. Therefore, when the interval duration varies from one to another (i.e., t_i not constant), Eq. 9 must be used

At the canopy scale, Δe depends on air vapor pressure and temperature of the canopy surface because leaf (actually leaf interior) vapor pressure is a function of leaf temperature. However, since canopy temperature is not generally available, the practical normalization of $WP_p{}^c$ and WP_b for evaporative demand (Tanner and Sinclair 1983) is usually approximated by replacing Δe with the atmospheric vapor pressure saturation deficit (VPD, using same units), implicitly assuming that air and canopy temperatures are the same.

This approximation has important drawbacks. For one, Δe could be substantially lower than VPD when canopy temperature is cooler than air temperature under high transpiration flux or positive advection of sensible heat, and Δe could be substantially higher than VPD when canopy temperature is hotter than air temperature due to water stress or negative advection of sensible heat (Paw U and Gao 1988; Asseng and Hsiao 2000). Furthermore, the normalization becomes very sensitive to low VPD values, giving unreliable results in such conditions (Stockle et al. 2003). In addition to the inaccuracy of the approximation under advective conditions, normalization by VPD encounters a more practical problem—a strong dependence of the calculated value of VPD on the type of weather data available and the method selected for calculation (Allen et al. 1998). Thus, the normalized WP may be substantially different unless the same kind of weather data (whether hourly, daily mean from hourly, daily based on maximum and minimum, etc.) and the same method of calculation are used.

Penman (1948), in his derivation of the latent heat flux equation named after him, found a better solution by approximating the canopy temperature with air temperature and the approximate slope of saturated vapor pressure versus temperature, and then combining the sensible heat flux equation and the energy balance equation. The solution given by Penman (1948) is the well known latent-heat flux equation that combines the "energy" and the "aerodynamic" components of the latent heat flux and that was further extended by Monteith (1980) to incorporate the aerodynamic and canopy components through conductance parameters. The fully developed Penman–Monteith equation is expressed as

$$\lambda E = \frac{s(Rn - G) + \rho_a c_p g_a (e^* - e_a)}{s + \gamma\left(1 + \frac{g_a}{g_c}\right)} \tag{10}$$

where Rn is the net radiation flux (J m^{-2} s^{-1}), G is the storage heat flux (J m^{-2} s^{-1}), g_a and g_c are the aerodynamic and canopy conductances (m s^{-1}), and e^* and e_a are, respectively, the saturation (at air temperature) and actual vapor pressure of the atmosphere. The other variables have already been defined. Note that although only VPD ($e^* - e_a$) appears in the equation, Eq. 10 actually does use vapor pressure at the approximated canopy temperature in arriving at the driving force for transpiration because sensible heat transfer is an integral though implicit part of the equation.

The more accurate modeling of the transpiration flux through the Penman–Monteith equation avoids the drawbacks of using VPD in place of Δe and leads to the second normalizing route, i.e., via a reference transpiration flux (E_o). In this case, the normalization of WP_p^c and WP_b are expressed respectively through the following equations:

$$\mathrm{WP}_{\mathrm{p}}^{\mathrm{c},\otimes} = \frac{\sum_{i=1}^{n} A_{\mathrm{c,i}}}{\sum_{i=1}^{n} \left(\frac{T_c}{E_o}\right)_i} \tag{11}$$

$$\mathrm{WP}_{\mathrm{b}}^{\otimes} = \frac{\mathrm{Biomass}}{\sum_{i=1}^{n} t_i \left(\frac{T}{E_o}\right)_i} \tag{12}$$

The summations are the same as those for Eqs. 7 and 8. A_c, T_c and E_o denote the quantity cumulated during each interval *i*. Again, for practical purposes an alternative form of Eq. 12 is often used, with Eo‾‾‾‾Eo‾being the mean daily E_o for the time interval *i*. With t_i varying, then the following equation must be used:

$$\mathrm{WP}_b^{\otimes} = \frac{\text{Biomass}}{\sum_{i=1}^{n}\left(\frac{T_c}{\overline{E_o}}\right)_i} \tag{13}$$

This approach was originally adopted by de Wit (1958), using pan evaporation in place of E_o. Nowadays, it is suggested to use the reference crop evapotranspiration as E_o and calculated according to the Penman–Monteith equation (Eq. 10), as rearranged and recommended by FAO, Irrigation and Drainage Paper no. 56 (Allen et al. 1998).

Various experimental works have shown the normalization of $\mathrm{WP_b}$ by E_o to be more robust than the normalization by VPD (e.g., Azam-Ali et al. 1994; Clover et al. 2001; Steduto and Albrizio 2005). After all, this approach has already been used very successfully for normalizing crop evapotranspiration, yielding the crop coefficients (K_c) used worldwide (Doorenbos and Pruit 1977). An example comparing normalization by VPD and by E_o is shown in Fig. 7a, b, respectively, for the same crops depicted in Fig. 4 through 6. Another example comparing the two ways of normalization is given in Fig. 8a, b. In these figures, it is easily seen that compared to normalization by VPD, normalization by E_o gave results that are more unified and hence should be more useful in applications across a range of conditions differing in evaporative demand.

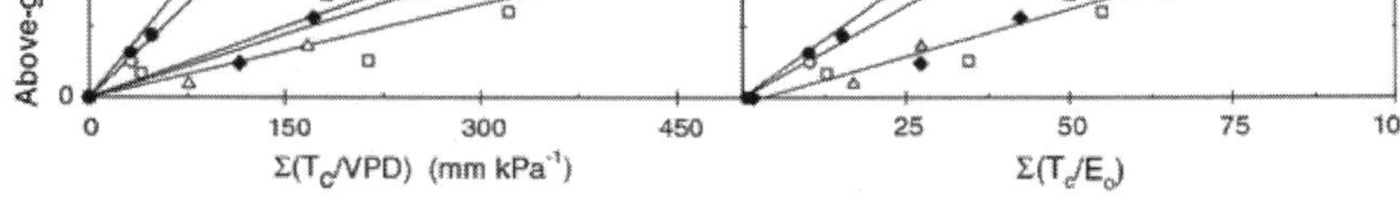

Figure 7. Relationship between aboveground biomass and cumulative crop transpiration when normalized **a** for day-time saturation vapor pressure deficit of the atmosphere (VPD) and **b** for daily reference-crop evapotranspiration (E_o). Data from the same sets of experiments as those depicted in Figs. 3 through 5, but with only pre-anthesis data for sunflower. Data refer to period from June to September, from May to August, from

March to May, and from April to June, for sorghum, sunflower, wheat and chickpea, respectively. Note the normalization for E_o caused the grouping of all C_3 species on a line with a single slope (13.4 g m^{-2}). Slopes for Sorghum are 25 and 32.9 g m^{-2} for the unfertilized (N_o) and fertilized (N_1) treatments, respectively. Redrawn from Steduto and Albrizio (2005)

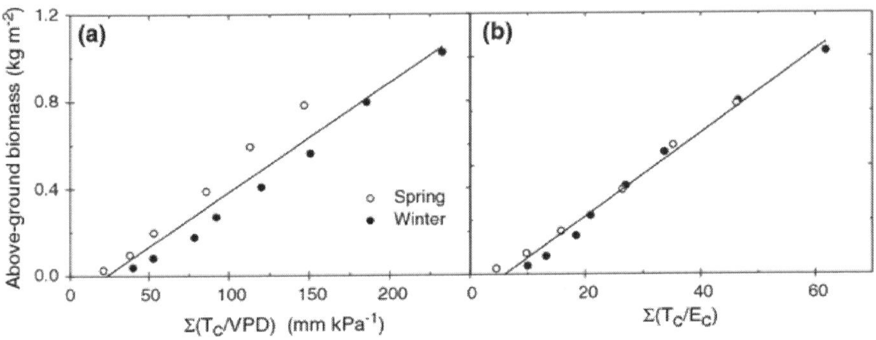

Figure 8. Relationship between aboveground biomass and cumulative crop transpiration: when normalized **a** for daily saturation vapor pressure deficit of the atmosphere (VPD) and **b** for reference-crop evapotranspiration (E_o). Data were obtained in Cordoba, Spain, on sunflower planted in the field either in winter (December–February) or spring (March–April). Note the normalization for E_o caused the grouping of the data for the winter and spring plantings to coalesce more along the regression line having a slope of 17.5 g m^{-2}. Derived from the original data of Soriano et al. (2004)

In our studies, the advantage of normalizing by E_o was evident when comparing data obtained on crops grown in atmospheric environments that differed in temperature, vapor pressure, and radiation regimes (i.e., winter vs. spring). When crops were grown in one location under similar environments but in different years, in one comparison there were no obvious differences between the two normalization procedures (T.C. Hsiao, unpublished).

Normalization for different atmospheric CO_2 concentrations

Similar to the WP_b normalization for different atmospheric evaporative demands, also the normalization of WP_b for different atmospheric CO_2 concentrations requires a reference status, in this case of CO_2concentration. This approach was alluded to by Tanner and Sinclair

(1983), and was more fully developed through the analysis by Hsiao (1993).

The conceptual points of departure are (1) the leaf gas-exchange processes formulation under steady-state conditions, as expressed by Eqs. 1 and 2, and (2) the accumulated evidence indicating a strong tendency of the ratio c_i/c_a (= α) to remain constant (Hsiao and Jackson 1999), attributed to the evolutionary adaptation of plants to environment and consistent with the theory of optimal stomatal behavior in water use (Cowan1982).

Expressing c_i as a function of c_a (i.e., $c_i = \alpha c_a$) and substituting in Eq. 3, yields

$$\mathrm{WP_p} = 0.625\,(1-\alpha)\,\frac{c}{\Delta w} \tag{14}$$

In view of the conservative behavior of α, it is advantageous to express the $\mathrm{WP_p}$ under a different CO_2concentration on a relative basis, i.e., as compared to its value under a reference CO_2 concentration. Indicating with subscript "o" the $\mathrm{WP_p}$ of the reference situation, the $\mathrm{WP_p}$ of any new situation relative to the reference situation is expressed as

$$\frac{\mathrm{WP_p}}{\mathrm{WP_{p,o}}} = \frac{0.625\,(1-\alpha)}{0.625\,(1-\alpha_o)}\,\frac{c_a}{\Delta w}\,\frac{\Delta w_o}{c_{a,o}} \tag{15}$$

That, after cancellation (with $\alpha \cong \alpha_o$), reduces to

$$\frac{\mathrm{WP_p}}{\mathrm{WP_{p,o}}} = \frac{c_a}{c_{a,o}}\,\frac{\Delta w_o}{\Delta w}$$
$$\therefore \quad \mathrm{WP_p} = \mathrm{WP_{p,o}}\,\frac{c_a}{c_{a,o}}\,\frac{\Delta w_o}{\Delta w} \tag{16}$$

Equation 16 indicates that $\mathrm{WP_p}$ for any situation is the product of the reference $\mathrm{WP_{p,o}}$, the c_a ratio, and the Δw ratio; and it represents the first paradigm equation (Hsiao 1993) for investigating changes in water productivity under different CO_2 concentrations (e.g., elevated future CO_2). It should also be recognized that, in order to normalize the value of $\mathrm{WP_p}$, it is necessary to know its reference value ($\mathrm{WP_{p,o}}$). Presently there is not yet a common reference c_a, and researchers are likely to choose a particular $\mathrm{WP_{p,o}}$ for their specific crop at their own convenience (e.g., Xu and Hsiao 2004). Progress in analyzing WP would be facilitated if $c_a$of a particular time (e.g., summer) and year (e.g., 2000) measured at a particular location (e.g., Mauna Loa Observatory, Hawaii) is chosen as a common reference.

Even though Eq. 15 was derived from equations describing the gas-exchange of single leaves, Xu and Hsiao (2004) have shown that the equation predicted quite well the hour-by-hour changes in $WP_p{}^c$ of a cotton canopy in the open field. Apparently, the scaling up from leaf to canopy did not require any modification of Eq. 15. This is likely the result of Eq. 15 expressing WP_p relative to $WP_{p,o}$, as up scaling from leaf to canopy should be identical or very similar for different situations regardless of environmental conditions (Xu and Hsiao 2004).

Although Eq. 16 remains valid when applied to canopies, as long as α stays the same, additional considerations are required when extending it to biomass. At aggregated canopy scale, biomass can be expressed as

$$\text{Biomass} = c\left(\sum_{i=1}^{n}(A_c)_i - \sum_{i=1}^{n}(R_c)_i\right) \tag{17}$$

where, C is the biomass per unit of carbon dioxide for a given composition; A_c is the net crop assimilation during day-time and R_c is the crop dark respiration during night-time; both A_c and R_c are integrated over the number of days (n) during which the biomass is produced.

Since the relationship between R_c and A_c tends to remain conservative, as previously indicated in "Photosynthetic water productivity—canopy scale" and "From net carbon gain to biomass", by setting β as the ratio of R_c/A_c, so that $R_c = \beta A_c$, and substituting into Eq. 17, one obtains the following:

$$\begin{aligned} \text{Biomass} &= C\left(\sum_{i=1}^{n}(A_c)_i - \beta\sum_{i=1}^{n}(A_c)_i\right) \\ \therefore\quad \text{Biomass} &= C\,(1-\beta)\sum_{i=1}^{n}(A_c)_i \end{aligned} \tag{18}$$

Using Eq. 18, the biomass water productivity (WP_b) can then be expressed as

$$WP_b = \frac{\text{Biomass}}{\sum_{i=1}^{n}(T_c)_i} = \frac{C\,(1-\beta)\sum_{i=1}^{n}(A)_i}{\sum_{i=1}^{n}(T)_i} \tag{19}$$

In view of the conservative behavior of β, also in this case it is advantageous to express the change of WP_b on a relative basis, i.e., its value under new CO_2 as compared to the reference value under the reference CO_2 concentration. Therefore, in analogy to Eq. 15, and using the same subscript and superscript symbols, we can derive

$$\frac{\mathrm{WP_b}}{\mathrm{WP_{b,o}}} = \frac{\mathrm{Biomass}}{\sum_{i=1}^{n}(T_{\mathrm{c}})_i}\frac{\sum_{i=1}^{n}(T_{\mathrm{c,o}})_i}{\mathrm{Biomass_o}}$$

$$\Rightarrow \frac{\mathrm{WP_b}}{\mathrm{WP_{b,o}}} = \frac{C(1-\beta)\sum_{i=1}^{n}(A_{\mathrm{c}})_i\sum_{i=1}^{n}(T_{\mathrm{c,o}})_i}{C(1-\beta_{\mathrm{o}})\sum_{i=1}^{n}(T_{\mathrm{c}})_i\sum_{i=1}^{n}(A_{\mathrm{c,o}})_i}$$

That, after cancellation (with $\beta_n \cong \beta_o$) and rearrangement, reduces to

$$\frac{\mathrm{WP_b}}{\mathrm{WP_{b,o}}} = \frac{\sum_{i=1}^{n}(A_{\mathrm{c}})_i}{\sum_{i=1}^{n}(A_{\mathrm{c,o}})_i}\frac{\sum_{i=1}^{n}(T_{,o})_i}{\sum_{i=1}^{n}(T)_i} = \frac{\sum_{i=1}^{n}(\mathrm{WP_p^c})_i}{\sum_{i=1}^{n}(\mathrm{WP_{p,o}^c})_i} \quad (20)$$

Equation 20 indicates that the relative change in WP_p is the product of the A_c ratio by T_c ratio, which corresponds to the $WP_p{}^c$ ratio of the gas-exchange at canopy level over the number of days (n).

Since, as mentioned before, the same equation applies when up-scaling from leaf to canopy gas-exchange, Eq. 20 can be rewritten in analogy to Eq. 16 as

$$\mathrm{WP_b} = \mathrm{WP_{b,o}}\frac{\sum_{i=1}^{n}(c_{\mathrm{a}})_i\sum_{i=1}^{n}(\Delta \mathrm{w_o})_i}{\sum_{i=1}^{n}(c_{\mathrm{a,o}})_i\sum_{i=1}^{n}(\Delta \mathrm{w})_i} \quad (21)$$

Equation 21 indicates that WP_b for any situation (e.g., elevated CO_2) is the product of the reference value ($WP_{b,o}$) times the c_a ratio and times the Δw ratio (as is for Eq. 16), cumulated over the same number of days during which the biomass is produced.

To normalize the biomass water productivity value under any atmospheric CO_2 concentration (of the past or of the future) to the reference value ($WP_{b,o}$) it is just a matter of rearranging the same equation as follows:

$$\mathrm{WP_{b,o}} = \mathrm{WP_b}\frac{\sum_{i=1}^{n}(c_{\mathrm{a,o}})_i\sum_{i=1}^{n}(\Delta \mathrm{w})_i}{\sum_{i=1}^{n}(c_{\mathrm{a}})_i\sum_{i=1}^{n}(\Delta \mathrm{w_o})_i} \quad (22)$$

Data with sufficient details to test the validity of Eqs. 21 and 22 are rare. In one study, cotton plants were grown in pots in controlled environment chambers under normal, 1.5 × normal and 2 × normal air CO_2concentrations and their WP_b was determined. Other than the differences in CO_2, the chambers were the same in environment. As expected, the results (Table 1) show that the measured WP_b was positively correlated with the level of c_a. Designating the normal air CO_2 concentration (360 ppm) as the reference situation, i.e., its c_a as $c_{a,o}$ and its Δw as Δw_o, the normalized WP_b ($WP_{b,o}$) was calculated for the two elevated CO_2 treatments. It is seen in the fifth column of

Table 1 that the resultant $WP_{b,o}$ was nearly identical in value compared to the measured WP_b for the reference situation. The WP_b of the elevated CO_2treatments was then predicted with Eq. 21 from the designated $WP_{b,o}$ (3.93 g kg^{-1}) the known c_a ratio, and the Δw ratio. The results (Table 1, last column) show that WP_b at elevated CO_2 predicted by Eq. 21 agreed closely with the measured values. The close agreement (within 2%) suggests strongly that Eqs. 15 and 21adequately account for all the important variables that affect biomass water productivity at least in cotton. Nonetheless, it is desirable to obtain more data of this nature to test further the validity of Eqs. 15, 21 and22, especially for other crop species.

Table 1. Measured biomass water productivity (WP_b), normalized biomass water productivity ($WP_{b,o}$), and predicted WP_b of cotton plants grown in pots under three levels of air CO_2 concentration in controlled environmental chambers

Nominal c_a (ppm)	r^2 for WP_b determination	$\Delta w/\Delta w_o$	WP_b measured (g kg^{-1})	$WP_{b,o}$ calculated (Eq. 22) (g kg^{-1})	WP_b predicted (Eq. 21) (g kg^{-1})
360	0.998	1	3.93	3.93	–
540	0.997	0.967	5.15	3.87	5.23
720	0.996	0.934	7.17	3.85	7.32

Measured WP_b was determined from plots of biomass produced versus cumulative transpiration by linear regression, and the coefficients of determination are given in the table as r^2. Designating the normal air CO_2 concentration (360 ppm) as the reference situation, biomass water productivity of the plants grown at either 540 or 720 ppm CO_2 were normalized using Eq. 22 with the c_a and Δw of the reference situation. Predicted WP_b was calculated using Eq. 21 with WP_b of the reference situation ($WP_{b,o}$ = 3.93 g kg^{-1}). Since chamber conditions were kept the same from day to day, summation called for in Eq. 21 and 22 was not necessary and the overall c_a and Δw were used in the calculation. Plants were grown over a 43-day period under about 700 µmol m^{-2} s^{-1} of PAR, 27/20°C and 45/80% relative humidity for the day and night periods. Periodically shoot biomass was harvested while water lost by transpiration for each period was measured by water balance. Soil surface was covered to eliminate evaporation. Air CO_2 was measured intermittently to obtain an overall c_a that was slightly different from the nominal set values of the chamber. Foliage temperature data used to calculate Δw were obtained in a separate experiment under very similar conditions. Derived from Hsiao and Xu (2000b)

Measured WP_b was determined from plots of biomass produced versus cumulative transpiration by linear regression, and the coefficients of determination are given in the table as r^2. Designating the normal air CO_2concentration (360 ppm) as the reference situation, biomass water productivity of the plants grown at either 540 or 720 ppm CO_2 were normalized using Eq. 22 with the c_a and Δw of the reference situation. Predicted WP_b was calculated using Eq. 21 with WP_b of the reference situation ($WP_{b,o}$ = 3.93 g kg^{-1}). Since chamber conditions were kept the same from day to day, summation called for in Eq. 21 and 22 was not necessary and the overall c_a and Δw were used in the calculation. Plants were grown over a 43-day period under about 700 µmol m^{-2} s^{-1} of PAR, 27/20°C and 45/80% relative humidity for the day and night periods. Periodically shoot biomass was harvested while water lost by transpiration for each period was measured by water balance. Soil surface was covered to eliminate evaporation. Air CO_2 was measured intermittently to obtain an overall c_a that was slightly different from the nominal set values of the chamber. Foliage temperature data used to calculate Δw were obtained in a separate experiment under very similar conditions. Derived from Hsiao and Xu (2000b)

Equations 16 and 21 are fundamental and should hold regardless of whether plants are C_3 or C_4, and under various sets of changes in environmental conditions (i.e., not only CO_2). This normalizing approach for different atmospheric CO_2 concentrations is expected to be the most valuable and robust one since it is based on the conservative behavior of α (c_i/c_a) and β (R_c/A_c), most likely the consequence of natural evolution and adaptation of plants in resources-use optimization.

The normalizations for evaporative demand and for atmospheric CO_2 have been treated separately, although Eq. 22 has the potential to be used for the purpose of integrated normalization. That is because it expresses $WP_{b,o}$ as a function of both the c_a ratio needed to normalize for CO_2 and the Δw ratio that is a part of Eqs. 7, 8, or 9 for the normalization for evaporative demand. Unfortunately in practice the data demanded by Eq. 22 are not readily available for crops grown in open fields or greenhouses. This problem will be elaborated on in "Uncertainties in the normalization for atmospheric CO_2", where a practical approach to normalize for the two key variables will be given.

Some important implications of normalizing biomass water productivity are clear. First, it allows the comparison of WP data across the globe on equal footing, after accounting for differences due to variations in evaporative demand of the climate, and in atmospheric carbon dioxide concentration when applicable. Such comparisons will reveal more definitively the intrinsic properties of the crop or the management practices that alter WP. Most importantly, normalized WP will provide a head start in knowing the WP at a new location or new time period when CO_2 concentration is different, whether in the future or in the past. For example, Eq. 21 will permit the prediction of WP_b once the CO_2 concentration for a situation is known and the vapor pressure gradient (Δw) is estimated.

SOME UNRESOLVED ISSUES

Up to this point we have focused on the basics and the apparently more straight forward relationships, even though there are possible complicating effects in some of our assumptions, and uncertainty over the degree of variation in α with respect to some environmental features. At least four issues deserve more attention and are briefly discussed here.

Effects of air humidity and water stress on α

The fact that stomata of many species open less under large Δw is well known (Schulze and Hall 1982). It has been shown (Morison 1987) that α changes in close association with stomatal changes caused by Δw. So as Δw changes dynamically in the open field, α would be expected to change dynamically as the result. Fortunately, the empirically determined relationship between α and Δw appears to be simple, with α decreasing nearly linearly as Δw increases (Morison 1987; Xu and Hsiao 2004). Thus, the pattern of α values could be estimated from the pattern of Δw using their linear empirical relationship. Xu and Hsiao (2004) have shown that by adjusting α according to Δw with the linear relationship, their theoretical model was able to predict the diurnal changes in canopy WP_p of cotton in the open field. Without the adjustment the prediction deviated markedly from the measured WP_p for a substantial part of the day. For the estimation of WP over the long term, however, how to obtain the most valid integration of α remains to be worked out.

Studies based on carbon isotope discrimination have shown (e.g., Johnson and Tieszen 1993) or inferred (e.g., Acevedo 1993) a reduction in α under water deficits. These results are probably confounded, however, by the likely stomatal response to Δw. Under conditions of water deficit, transpiration is lower due to reduced stomatal conductance; consequently the foliage is warmer and Δw is higher. The reduced α under water deficit may be the result of this change in Δw, and not a direct response to water deficit. In addition, much of the isotope data were obtained by stressing the plants in pot cultures where stress develops much faster than in the field. Slow adjustment to stress appears to be needed for α to remain constant.

Genotypic variation in α

The evidence for differences in α among genotypes is mostly based on carbon isotope discrimination (Farquhar et al. 1989), a measure widely used in efforts to breed for higher WP and resistance to drought (see: Acevedo 1993; Wright et al. 1993; Condon et al. 2004, for review). Different genotypes of various C_3crop species exhibited under similar conditions only limited differences (e.g., 10%) in α when well watered. The probable exceptions are genotypes of peanut, which exhibited apparently larger differences in α as inferred from carbon isotope discrimination data (Wright et al. 1993). The significant efforts devoted to breeding for lower α and higher WP have been constrained because, in most cases, less isotope discrimination (lower α) is associated with slower biomass production and lower yield when water supply is not severely limiting (Acevedo 1993; Condon et al. 2004). The work of Rebetzke et al. (2002) in wheat demonstrates the extensive efforts necessary to make only marginal yield improvements under water deficit conditions using carbon isotope discrimination as a selection tool.

In all cases described above where α has values significantly different from the values of 0.7 for C_3 and 0.4 for C_4 noted in "Photosynthetic water productivity—leaf scale", but not changing dynamically, it is simply a matter of using the correct α value in Eq. 5 in making assessments of WP. In cases of normalizing for different CO_2 concentrations using Eq. 22, or its equivalent rearrangement of Eq. 16, if α is significantly different between the reference situation and the actual situation, it is necessary to substitute $(1 - \alpha)c_a$ in place of c_a, for both situations.

Effects of nitrogen nutrition and changing shoot-root ratio on the constancy of WP_b

Two other situations may introduce some variation in WP. On the one hand, it has been shown that α remains nearly constant over a range of N nutrition (Wong et al. 1979; Fig. 2). Nevertheless, when the N deficiency is severe, the decrease in leaf photosynthesis has been associated with a decrease in WP_b. Crop plants grown under limited N adjust primarily their leaf area before their intrinsic photosynthetic rate is affected (Sinclair and Horie 1989). Thus, one would expect that WP_b would tend to decrease in low N supply situations only in cases of quite severe N deficiency.

The known response of a shift in assimilate partitioning toward the root system relative to the shoot caused by water deficits (Hsiao and Xu 2000b) could induce an apparent change in WP_b in situations where water stress increases in severity with time. This is because a larger and larger proportion of the assimilated carbon would be going to the roots and not to the shoot, the part of the biomass normally harvested to measure WP_b. The maintenance of linearity over a substantial range of ET deficits in the myriad of published biomass-ET relations suggests, however, that this effect may not have a significant influence until water deficits become quite severe.

Uncertainties in the normalization for atmospheric CO_2

Still other uncertainties are in the normalization for c_a. To normalize with Eqs. 15, 16 or 22, the ratio of CO_2as well as the ratio of water vapor concentration are required as input. Although the former may be taken as the measured value at the reference station or even the common value estimated for the world's atmosphere without much error, the latter ($\Delta w/\Delta w_o$) is not simple to assess. Of course w_a is to be taken from weather data, but w_i and hence Δw is a function of canopy temperature. Canopy temperature is best obtained by measurements; but data of measured canopy temperature over the weather dynamics of a growing season are rare. Canopy temperature can be estimated from weather data using the Penman–Monteith equation (Eq. 10), but detailed data on radiation, air temperature and humidity, and wind velocity, as well as canopy conductance (g_c) are required. At the present, g_c has been parameterized for use in the Penman–Monteith equation by Allen et al. (1998) to be constant at 14.3 mm s^{-1} (corresponding to 70 s m^{-1} in terms of resistance). Response of stomata of single leaves to changes in c_a is well known and has been characterized for a number of species (Morison 1987), but how changes in stomatal conductance scale up to changes in canopy conductance is still an open question (e.g., Baldocchi et al. 1991; Rochette et al. 1991), and well quantified data to facilitate the resolution are rare. The only certainty is that g_c would decrease with increases in atmospheric CO_2.

In view of the reality that long time $\Delta w/\Delta w_o$ data are not simple to obtain in the future and mostly none-existing in the past, for practical purposes it is recommended for now that the following semi-empirical equation be used to normalize WP for atmospheric CO_2:

where $c_{a,o}$ and c_a are the annual mean atmospheric CO_2 concentrations measured at Mauna Loa Observatory (Hawaii), respectively for the reference year and for the year when WP_b is determined; and D is an empirical factor approximating the sum of Δw to the sum of Δw_o ratio (Eq. 22). In theory and according to the $\Delta w/\Delta w_o$ data in Table 1, D should decrease with increases in c_a, and the decrease should be slight. As shown in Table 1, the ratio appears to be reduced about 7% even for the case of a doubling in CO_2. The data are for cotton grown in environmental chambers, where air turbulence (hence boundary layer conductance for sensible heat, which relates to canopy temperature) is less than in the open field. Therefore the Δw ratio for cotton growing in the field under a doubling of CO_2 should be even closer to 1.0. For now as a temporary measure, it is suggested that D be approximated by the equation $D = a - b(c_a - c_{a,o})$, where $a =$ 1.0, and $b=$ 0.000138. This relationship is based on the $\Delta w/\Delta w_o$ data of Table 1 and assumes the Δw ratio to be a linear function of the difference in CO_2 concentration between the given and the reference situation. It also assumes, arbitrarily, that the effect of difference in CO_2 on the ratio is reduced by 25% in the open field due to more turbulence. Implicit in the equation is the fact that the $c_{a,o}$ is taken to be 360 ppm. If the chosen $c_{a,o}$ is substantially higher or lower, the b parameter should be reevaluated. Obviously, Eq. 23 can be rearranged to predict WP_p at different c_a.

It is emphasized that Eq. 23 can only be used to normalize for CO_2 and does not account for any difference caused by evaporative demand of the air. That is because the D factor is based on data obtained under the same environmental conditions in the growth chambers where air evaporative demand was constant. Consequently, for now it is recommended that biomass water productivity should first be normalized for atmospheric CO_2 using Eq. 23, then the result be further normalized for evaporative demand using Eq. 13 (or Eq. 12). It is clear from this discussion that field data on canopy conductance and $\Delta w/\Delta w_o$ under different air CO_2 concentration and for different crops are badly needed for the development of a better CO_2normalization procedure in the future.

Yet, another practical question is how to settle on a reference state needed for CO_2 normalization. Taking as a given that a common reference year or $c_{a,o}$ is agreed upon by the research community, there is still the question of agreeing on the value of $WP_{b,o}$ of the given crop species, and on the associated Δw_o, which is partly dependent on the specific weather record over the growing season. Obvious this can only be done by an organized effort, and substantial lead time is necessary to analyze the literature and developing new data before a unified base can be developed. As of now, it is likely that most researchers will choose the reference situation for their own convenience.

CONCLUSIONS

The above stepwise approach, from leaf to the whole canopy, has provided a conceptual and theoretical framework to explain the basis for the constancy of biomass water productivity (WP_b). Following the pioneer work of de Wit (1958), Bierhuizen and Slatyer (1965), and Tanner and Sinclair (1983), the approach here added some new perspectives and new kinds of supporting data. Important is the fact that the scaling up does not involve complicated models and is underpinned at each level by theory and fundamentals, and by relevant data either recently published or new. Progress was made in the normalization for the two key climatic variables, evaporative demand and atmospheric CO_2 concentration. We have justified by theory the use of reference ET (E_o) in place of air VPD for the evaporative demand normalization, and shown it to be superior by some experimental data. Equally significant is the normalization for atmospheric CO_2concentration, given its past changes and the certainty of its increase in the future. We have shown that, at the leaf level, because of the coordination of photosynthetic capacity and stomatal opening to maintain $c_i$nearly constant relative to c_a, the normalized WP becomes a function of the external CO_2 concentration. The CO_2 normalization approach, tested with limited data to be also applicable at the canopy level, allows for the assessment of crop production under future climate scenarios.

As for temporal scales, canopy photosynthetic WP, though varying on a diurnal basis due to variations in evaporative demand and air CO_2 concentration (Xu and Hsiao 2004), remains conservative when averaged over periods of days up to a season. Apparently the tight relation between photosynthesis and respiration contributed to that conservative behavior. The transition from photosynthetic to biomass WP depends on biomass composition. While the metabolic costs of biosynthesis do not vary significantly among crop species for vegetative structures, except for the difference between C_3 and C_4, it is strongly dependent on the protein and lipid content of fruits and seeds. Thus, normalized WP_b should not differ much for crops of similar composition, although its value should decrease from cereals, to legumes, to oil crops. Recent experimental evidence presented above provides support for the theory and illustrates the conservativeness of normalized WP_b under a wide variety of conditions.

Such features of normalized WP_b offer an invaluable tool for modeling crop production as related to water supply and availability, providing an effective way of applying the normalized WP_b values between different locations, climate and seasons. The robustness of WP_b, within the limits and uncertainties discussed above, paves the way for its use as the pivotal function driving the quantitative assessment of water-limited productivity. Another conservative parameter used for modeling crop productivity is radiation use efficiency (RUE). RUE, however, can remain variable even after normalizing for evaporative demand (Sinclair and Muchow 1999; Albrizio and Steduto 2005) and gives less consistent results in comparison to WP_b in some cases (Steduto and Albrizio 2005).

An alternative use of WP_b can be for the mapping of climatic zones as different classes of biomass water productivity and given cropping systems so that optimal spatial allocation could be developed to conserve the limited water resources. Moreover, WP_b can be used as a simple measure of consumptive water use by crops. This can be achieved by dividing the weight of biomass (as dry matter) sampled from a cropped field by the pertinent WP_b value, resulting in the crop transpiration cumulated up to the sampling time. In conjunction with estimates of soil evaporation, the overall cumulative evapotranspiration can be estimated as well. However, more field research is needed to broaden the application of constant WP_b, particularly under more extreme stresses, various rates of stress development, different level of mineral nutrition and air CO_2 concentration, and diverse VPD and temperature regimes. Additionally, more extensive determination of WP_b for crops having different biomass composition is needed.

As elaborated on in this paper, it is obvious that WP_b of crops, being conservative, offers only limited opportunity for improving the productivity of the water consumed by crops. As briefly discussed, WP_b can be raised by growing the crop under weather regimes of lower evaporative demand or by improving nitrogen nutrition. The potential improvements outside drastic modification of the environment (e.g., greenhouses), however, are quite limited. Raising CO_2 concentration of the air offers a greater potential in theory, and can be achieved with current technology in protected cultivation such as in plastic tunnels and glasshouses. For production in the open field, enrichment with CO_2 of the air surrounding the crop, though tested from time to time, is not a foreseeable practical possibility. On the other hand, the current relentless rise in atmospheric CO_2 with time is certainly to raise WP_b of crops in the future. For example, 50% increase in c_a would raise WP_b by slightly less than 50% according to Eq. 21.

Given the conservative behavior of WP_b, and its relatively narrow range of variation between genotypes, it seems that promising research directions for its improvement can come only from some genetic breakthroughs which would change the intrinsic carboxylation and respiration capacities of plants. Such breakthroughs are extremely difficult to achieve since the natural evolution of species has apparently already optimized complex adaptation mechanisms to natural resource use. In any case, the time frame for such breakthroughs to occur would be quite long, in decades at least.

In considering the overall productivity of water, the exchange of transpirational water for biomass production (WP_b) is only one of the many sequential steps starting from water supplied either as rain or as delivered water leading to the final crop yield. The other steps encompass the engineering and management aspects of water storage, conveyance and irrigation application, water retention in the soil root zone and extraction by the crop, soil evaporation, and the partition of biomass between the vegetative residue and the yield part of the crop. Though WP_b is conservative, the other steps offer many possibilities to improve the overall water productivity. This is the subject of a companion paper in this issue (Hsiao et al. 2007).

ACKNOWLEDGEMENT

The authors thank the Land and Water Division of FAO, United Nations, for providing support in the research and preparation of this paper. The work of TCH has been partly funded by the US Department of Energy grant De-FG03-93-ER-6187, US Department of Agriculture NRICGP grant 99-35306-7793, and the CGIAR Water and Food Challenging Program through ICARDA. The authors are also grateful to Dr. Rossella Albrizio for sharing most of her experimental dataset. The help of Dr. Yoshiko Kasugi in obtaining and analyzing the data in Table 1 is much appreciated.

REFERENCES

1. Acevedo E (1993) Potential of carbon isotope discrimination as a selection criterion in barley breeding. In: Ehleringer JR, Hall AE, Farquhar GD (eds) Stable isotopes and plant carbon-water relations. Academic, San Diego, pp 399–417
2. Albrizio R, Steduto P (2003) Photosynthesis, respiration and conservative carbon use efficiency of four field grown crops. Agric For Meteorol 116:19–36
3. Albrizio R, Steduto P (2005) Resource use efficiency of field-grown sunflower, sorghum, wheat and chickpea. I. Radiation use efficiency. Agric For Meteorol 130:254–268
4. Allen RG, Pereira LS, Raes D, Smith M (1998) Crop evapotranspiration. Guidelines for computing crop water requirements (Irrigation and drainage paper 56). Food and Agriculture Organization, Rome
5. Amthor JS (1989) Respiration and crop productivity. Springer, New York, 215 p
6. Amthor JS (1995) Higher plant respiration and its relationships with Photosynthesis. In: Schulze ED, Caldwell MM (eds) Ecophysiology of photosynthesis. Springer, Berlin, pp 71–101
7. Angus JF, Hasegawa S, Hsiao TC, Liboon SP, Zandstra HG (1983) The water balance of post-monsoonal dryland crops. J Agric Sci (Camb) 101:699–710
8. Arkley J (1963) Relationships between plant growth and transpiration. Hilgardia 34:559–584
9. Asseng S, Hsiao TC (2000) Canopy CO_2 assimilation, energy balance, and water use efficiency of an alfalfa crop before and after cutting. Field Crops Res 67:191–206
10. Azam-Ali SN, Crout NMJ, Bradley RG (1994) Perspectives in modelling resource capture by crops. In: Monteith JL, Unsworth MH, Scott RK (eds) Resource capture by crops. Proceedings of the 52nd University of Nottingham Eastern School. Nottingham University Press, Nottingham, pp 125–148

11. Baldocchi DD, Luxmoore RJ, Hatfield JL (1991) Discerning the forest from the trees: an essay on scaling canopy conductance. Agric For Meteorol 54:197–226

12. Bierhuizen JF Slatyer OR (1965) Effect of atmosphere concentration of water vapor and CO_2 in determining transpiration-photosynthesis relationships of cotton leaves. Agric Meteorl 2:259–270

13. Biscoe PV, Scott RK, Monteith JL (1975) Barley and its environment. III. Carbon budget of the stand. J Appl Ecol 12:269–293

14. Björkman O (1981) The response of photosynthesis to temperature. In: Grace J, Ford ED, Jarvis PG (eds) Plants and their atmospheric environment. Blackwell, Oxford, London, Edinburgh, Boston, Melbourne, pp 273–301

15. Bolaños JA, Hsiao TC (1991) Photosynthesis and respiratory chracterization of field grown tomato. Photosynth Res 28:21–32

16. Briggs IJ, Shantz HL (1913a) The water requirement of plants: I. Investigation in the Great Plains in 1910 and 1911. USDA Bureau Plant Industry Bull, 284 p

17. Briggs IJ, Shantz HL (1913b) The water requirement of plants: II. A review of the literature. USDA Bureau Plant Industry Bull, 285 p

18. Brouwer R (1983) Functional equilibrium: sense or nonsense? Neth J Agric Sci 31:335–348

19. Charles-Edwards DA (1982) Physiological determinants of crop growth. Academic, Sydney, 158 p

20. Cheng W, Sims DA, Luo Y, Coleman JS, Johnson DW (2000) Photosynthesis, respiration and net primary production of sunflower stands in ambient and elevated atmospheric CO2 concentrations: an invariant NPP:GPP ratio? Glob Change Biol 6:931–941

21. Clover GRG, Jaggard KW, Smith HG, Azam-Ali SN (2001) The use of radiation interception and transpiration to predict the yield of healthy, droughted and virus-infected sugar beet. J Agric Sci 136:169–178

22. Condon AG, Richards RA, Rebetzke GJ, Farquhar GD (2004) Breeding for high water-use efficiency. J Exp Bot 55:2447–2460

23. Cowan IR (1982) Regulation of water use in relation to carbon gain in higher plants. In: Lange OL, Nobel PS, Osmod CB, Zigler H (eds)

Physiological plant ecology II. Ecycl plant physiol, NS, vol 12 B. Springer, Berlin, pp 589–613

24. Doorenbos J, Pruit WO (1977) Crop water requirements (Irrigation and drainage paper 24). Food and Agriculture Organization, Rome, 144 p
25. Farquhar GD, Richards RA (1984) Isotopic composition of plant carbon correlates with water-use efficiency of wheat genotypes. Aust J Plant Physiol 11:539–552
26. Farquhar GD, Sharkey TD (1982) Stomatal conductance and photosynthesis. Ann Rev Plant Physiol 33:317–345
27. Farquhar GD, Ehleringer JR, Hubick KT (1989) Carbon isotope discrimination and photosynthesis. Ann Rev Plant Physiol Plant Mol Biol 40:503–537
28. Fereres E, Goldhamer DA, Parson LR (2003) Irrigation water management of horticultural crops. HortScience 38(5):1036–1042
29. Garcia R, Kanemasu ET, Blad BL, Bauer A, Hatfield JL, Major DJ, Reginato RJ, Hubbard KG (1988) Interception and use efficiency of light in winter wheat under different nitrogen regimes. Agric For Meteorol 44:175–186
30. Gifford RM (1995) Whole plant respiration and photosynthesis of wheat under increased CO2 concentration and temperature: long-term vs. short-term distinctions for modelling. Glob Change Biol 1:385–396
31. Hanks RJ (1974) Model for predicting as influenced by water use. Agron J 66:660–665
32. Hanks JR (1983) Yield and water-use relationships: an overview. In: Taylor HM, Jordan WA, Sinclair TR (eds) Limitations to efficient water use in crop production. ASA, Madison, pp 393–411
33. Hirasawa T, Wakabayashi K, Touya S, Ishihara K (1995) Stomatal response to water deficits and abscisic acid in leaves of sunflower plants (*Helianthus annuus* L.) grown under different conditions. Plant Cell Physiol 36:955–964
34. Hsiao TC (1993) Effects of drought and elevated CO_2 on plant water use efficiency and productivity. In Jackson MD, Black CR (eds) Global environmental change. Interacting stresses on plants in a changing climate. NATO ASI series. Springer, New York, pp 435–465

35. Hsiao TC, Bradford KJ (1983) Physiological consequences of cellular water deficits. In: Taylor HM, Jordan WA, Sinclair TR (eds) Limitations to efficient water use in crop production. ASA, Madison, pp 227–265

36. Hsiao TC, Jackson RB (1999) Interactive effects of water stress and elevated CO_2 on growth, photosynthesis, and water use efficiency. In: Luo Y, Mooney HA (eds) Carbon dioxide and environmental stress. Academic, New York, pp 3–31

37. Hsiao TC, Xu L-K (2000a) Sensitivity of growth of roots vs. leaves to water stress: biophysical analysis and relation to water transport. J Exp Bot 51:1595–1616

38. Hsiao TC, Xu L-K (2000b). Predicting water use efficiency of crops. Acta Hortucult 537:199–206

39. Hsiao TC, Steduto P, Fereres E (2007) A systematic and quantitative approach to improve water use efficiency in agriculture. Irrig Sci (this issue)

40. Johnson RC, Tieszen LL (1993) Carbon isotope discrimination, water relations, and gas exchange in temperate grass species and accessions. In: Ehleringer JR, Hall AE, Farquhar GD (eds) Stable isotopes and plant carbon-water relations. Academic, San Diego, pp 281–296

41. Lambers H (1987) Growth, respiration, exudation and symbiotic association: the fate of carbon tranlocated to roots. In: Gregory PJ, Lake JV, Rose DA (eds) Root development and function. Cambridge University Press, London, pp 123–145

42. Landsberg JJ, Waring RH, Coops NC (2003) Performance of the forest productivity model 3-PG applied to a wide range of forest types. For Ecol Manage 172:199–214

43. Lhomme JP (1991) The concept of canopy resistance: historical survey and comparison of different approaches. Agric For Meteorol 54:227–240

44. Louwerse W (1980) Effects of CO_2 concentration and irradiance on the stomatal behavior of maize, barley, and sunflower plants in the field. Plant Cell Environ 3:391–398

45. Meek DW, Hatfield JL, Howell TA, Idso SB, Reginato RJ (1984) A generalized relationship between photosynthetically active radiation and solar radiation. Agron J 76:939–945

46. Monteith JL (1977) Climate and the efficiency of crop production in Britain. Philos T R Soc Lond B 281:277–294

47. Monteith JL (1980) The deveopment and extension of Penmans' evaporation formula. In: Hillel D (ed) Application of soil physics. Academic, New York, pp 247–253

48. Monteith JL (1984) Consistency and convenience in the choice of units for agricultural science. Exp Agric 20:105–117

49. Monteith JL (1993) The exchange of water and carbon by crops in a mediterranean climate. Irrig Sci 14:85–91

50. Morison JIL (1987) Intercellular CO_2 concentration and stomatal response to CO_2. In: Zeiger E, Farquhar GD, Cowan IR (eds) Stomatal function. Stanford University Press, Stanford, pp 229–251

51. Norman JM (1979) Modelling the complete crop canopy. In: Barfield BJ, Gerber JF (eds) Modification of the aerial environment of crops. ASAE, St. Joseph, pp 249–277

52. Osmond CB, Winter K, Ziegler H (1982) Functional significance of different pathways of CO_2 fixation in photosynthesis. In: Lange OL, Nobel PS, Osmond CB, Zeigler H (eds) Physiological plant ecology II. Water relations and carbon assimilation. Encyclopedia of plant physiology, new series 12B. Springer, Berlin, pp 479–547

53. Oweis T, Zhang H, Pala M (2000) Water use efficiency of rainfed and irrigated bread wheat in a Mediterranean environment. Agron J 92(2):231–238

54. Paw U KT, Gao W (1988) Applications of solutions to non-linear energy budget equations. Agric For Meteorol 43:121–145

55. Pearcy RW (1983) Physiological consequences of cellular water deficits: non-stomatal inhibition of photosynthesis by water stress. In: Taylor HM, Jordan WA, Sinclair TR (eds) Limitations to efficient water use in crop production. ASA, Madison, pp 277–287

56. Penman HL (1948) Natural evaporation from open water, bare soil and grass. Proc R Soc Lond Ser A 193:120–146

57. Penning de Vries FWT, Brunsting AHM, van Laar HH (1974) Products, requirements and efficiency of biosynthesis: a quantitative approach. J Theor Biol 45:339–377

58. Penning de Vries FWT, Van Laar HH, Chardon MCM (1983) Bioenergetics of growth of seeds, fruits, and storage organs. In: Potential productivity of field crops under different environments. International Rice Research Institute, Manila, pp 37–60

59. Rebetzke GJ, Condon AG, Richards RA, Farquhar GD (2002) Selection for reduced carbon isotope discrimination increases

aerial biomass and grain yield of rainfed bread wheat. Crop Sci 42:739–745

60. Rochette P, Pattey E, Desjardins RL, Dwyer LM, Stewart DW, Dube PA (1991) Estimation of maize (*Zea mays L.*) canopy conductance by scaling up leaf stomatal conductance. Agric For Meteorol 54:241–261

61. Schulze DE, Hall AE (1982) Stomatal responses, water loss and CO_2 assimilation rates of plant. In: Lange OL, Nobel PS, Osmond CB, Ziegler H (eds) Physiological plant ecology. II Encyclopedia of plant physiology, 12B. Springer, Berlin, pp 181–230

62. Seckler D, Amarasinghe U, Molden D, De Silva R, Barker R (1998) World water demand and supply, 1990 to 2025: scenarios and issues. International Water Management Institute Research Report 19, Colombo, Sri Lanka

63. Sinclair TR, Horie T (1989) Leaf nitrogen, photosynthesis, and crop radiation use efficiency: a review. Crop Sci 29:90–98

64. Sinclair TR, Muchow RC (1999) Radiation use efficiency. Adv Agron 65:215–265

65. Soriano MA, Orgaz F, Villalobos FJ, Fereres E (2004) Efficiency of water use of early plantings of sunflower. Eur J Agron 21:465–476

66. Stanhill G (1981) The size and significance of differences in the radiation balance of plants and plant communities. In: Grace J, Ford EP, Jarvis PJ (eds) Plant and their atmospheric environment. Balckwell, Oxford, pp 57–73

67. Stanhill G (1986) Water use efficiency. Adv Agron 39:53–85

68. Steduto P (1996) Water use efficiency. In: Pereira LS, Feddes RA, Gilley JR, Lesaffre B (eds) Sustainability of irrigated agriculture. NATO ASI series E: applied sciences. Kluwer, Dordrecht, pp 193–209

69. Steduto P, Albrizio R (2005) Resource-use efficiency of field-grown sunflower, sorghum, wheat and chickpea. II Water use efficiency and comparison with radiation use efficiency. Agric For Meteorol 130:269–281

70. Steduto P, Katerji N, Puertos-Molina H, Ünlü M, Mastrosilli M, Rana G (1997) Water use efficiency of sweet sorghum under water stress conditions. Gas-exchange investigations at leaf and canopy scales. Field Crops Res 54:221–234

71. Steduto P, Albrizio R, Giorio P, Sorrentino G (2000) Gas-exchange response and stomatal and non-stomatal limitations to carbon assimilation of sunflower under salinity. Environ Exp Bot 44:243–255

72. Stewart JI (1972) Prediction of water production functions and associated irrigation programs to minimize crop yield and profit losses due to limited water. PhD Thesis, University of California Davis. University Microfils#73–16, 934 p

73. Stöckle CO, Donatelli M, Nelson R (2003) CropSyst, a cropping systems simulation model. Eur J Agron 18:289–307

74. Tanner CB (1981) Transpiration efficiency of potato. Agron J 73:59–64

75. Tanner CB, Sinclair TR (1983) Efficient water use in crop production: research or re-search? In: Taylor HM, Jordan WA, Sinclair TR (eds) Limitations to efficient water use in crop production. ASA, Madison, pp 1–27

76. Taylor HM (1983) Managing root systems for efficient water use: an overview. In: Taylor HM, Jordan WA, Sinclair TR (eds) Limitations to efficient water use in crop production. ASA, Madison, pp 87–113

77. Varlet-Grancher C, Gosse G, Chartier M, Sinoquet H, Bonhomme R, Allirand JM (1989) Mis ou point: rayonnement solaire absorbé ou intercepté par un couvert végétal. Agronomie 9:419–439

78. Von Caemmerer S, Farquhar GD (1981) Some relationships between the biochemistry of photosynthesis and the gas exchange of leaves. Planta 153:376–387

79. Whitfield DM, Connor DJ, Hall AJ (1989) Carbon dioxide balance of sunflower subjected to water stress during grain-filling. Filed Crops Res 20:65–80

80. de Wit CT (1958) Transpiration and crop yields. Versl. Landbouwk. Onderz. 64.6 Inst. Biol. Chem. Res. On Field Crops and Herbage, Wageningen, The Netherlands

81. de Wit CT, Brouwer R, and Penning de Vries FWT (1970) The simulation of photosynthetic systems. In: Setlik (ed) Prediction and measurement of photosynthetic productivity. Pudoc, Wageningen, pp 47–70

82. Wong SC, Cowan IR, Farquhar GD (1979) Stomatal conductance correlates with photosynthetic capacity. Nature 282:424–426

83. Wright GC, Hubick KT, Farquhar GD, Rao RCN (1993) Genetic and environmental variation in transpiration efficiency and its correlation with carbon isotope discrimination and specific leaf area in peanut. In: Ehleringer JR, Hall AE, Farquhar GD (eds) Stable isotopes and plant carbon-water relations. Academic, San Diego, pp 247–267

84. Xu L-K, Hsiao TC (2004) Predicted vs. measured photosynthetic water use efficiency of crops stands under dynamically changing field environments. J Exp Bot 55:2395–2411

9

CHAPTER

Dynamics and Activity of Iron-Reducing Bacterial Populations in a West African Rice Paddy Soil under Subsurface Drainage: Case Study of Kamboinse in Burkina Faso

Cécile Harmonie Otoidobiga[1*], Amadou Keita[2], Hamma Yacouba[2], Alfred S. Traore[1], Dayéri Dianou[3]

[1]*Research Center for Biological, Alimentary and Nutritional Sciences, Research and Training Unit, Life and Earth Sciences, University of Ouagadougou, Ouagadougou, Burkina Faso*

[2]*International Institute for Water and Environmental Engineering, Ouagadougou, Burkina Faso*

[3]*National Center for Sciences and Technology Research, Ouagadougou, Burkina Faso*

ABSTRACT

Iron toxicity is one of the main edaphic constraints that hamper rice production in West African savanna and forest lowlands. Although chemical reduction processes of various types of pedogenic iron oxides could not be underestimated, the bulk of these processes can be ascribed to the specific activity of Iron-Reducing Bacteria (IRB). The reducing conditions of waterlogged lowland soils boost iron toxicity through the reduction of almost all iron into ferrous form (Fe^{2+}), which can cause disorder in rice plant and crop yield losses. Aiming to contribute at the improvement of rice yield in Africa, an experiment was developed to evaluate the impact of subsurface drainage on IRB dynamics and activity during rice cultivation. Twelve concrete microplots with a clay-loam soil and a rice variety susceptible to iron toxicity (FKR 19) were used for the experiment. Soil in microplots was drained for 7 days (P1), 14 days (P2), and 21 days (P3), respectively. Control (T) microplots without drainage were prepared similarly. The evolution of IRB populations and the content of ferrous iron in the paddy soil and in soil near rice root were monitored throughout the cultural cycle using MPN and colorimetric methods, respectively. Data obtained were analyzed in relation to drainage frequency, rice growth stage, and rice yield using the Student t test and XLSTAT 7.5.2 statistical software. From the results obtained, the subsurface drainage reduced significantly IRB populations ($p = 0.024$). However, the drainage did not affect significantly ferrous iron concentration in the soil near rice roots ($p = 0.708$). The concentration of ferrous iron ($p < 0.0001$) in soil near rice roots and the number of IRB ($p < 0.0001$) were significantly higher during the rice tillering and maturity stages. Although no significant difference was observed for rice yield among treatments ($p = 0.209$), the P2 subsurface drainage showed the highest yield and the lowest concentration of ferrous iron in soil near rice roots.

Keywords: Iron-Reducing Bacteria, Rice, Iron Toxicity, Subsurface Drainage

1. INTRODUCTION

Iron (Fe) toxicity is a nutritional disorder associated with high iron concentrations in the soil solution that affects rice plant in lowland [1] [2] . Large concentrations of ferrous Fe in soil solution may occur either when Fe is mobilized in situ by microbial reduction of ferric Fe [3] or when reduced Fe is translocated into valley bottoms by interflow or subsurface flow from adjacent slopes [4] . Iron toxicity affects more than 50% of lowlands in West Africa [1] [5] . In Burkina Faso, many lowland crop fields were even abandoned due to iron toxicity [1] [6] .

Iron toxicity is associated with large concentrations of reduced iron (Fe^{2+}) in the soil solution. As soon as a soil becomes flooded, the dissolved oxygen is consumed by aerobic bacteria and hemical oxidation reactions. Oxygen is depleted fast in most regions of the soil and alternative electron acceptors are used [7] . In anoxic conditions, Fe (III)-compounds are reduced and it results in ferrous ions production. This reduction of Fe(III) is largely controlled by microbial processes, and coupled to the oxidation of organic matter to support growth [8] . The appearance of iron toxicity symptoms in rice involves an excessive uptake of Fe^{2+} by the rice roots and its acropetal translocation into the leaves. A high production of toxic oxygen radicals can damage cell structural components and impair physiological processes [9] [10] . Excessive amounts of iron in the soil in reduced conditions upset the nutrient balance in plants [11] . Besides standard symptoms, i.e. browning and yellowing of leaves [9] , iron toxicity can cause a 10% - 100% crop yield losses [5] [12] .

The occurrence of Fe toxicity in flooded soils is often associated with poor drainage, poor nutrient status, acid pH reactions, and a low cation-exchange capacity of the soil [2] [13] .

Indeed, the poor level of development of lowlands in West Africa, epitomized with bad water management, generates favorable conditions for the occurrence of iron toxicity [14] - [16] . Thus, areas subjected to iron toxicity generally have drainage problems, and certain management practices, i.e. periodic soil drainage which allows oxidation of the root zone, can reduce iron toxicity for plant [17] [18] .

To contribute to the general effort to improve rice yield in Africa for food security, the present experiment was developed to measure the effect of subsurface drainage on microbiological and chemical parameters sustaining iron toxicity in paddy fields and on rice yield. To eliminate excess iron in soil through water management, concrete microplots were constructed, filled with a clay-loam soil and water was drained periodically during rice cultivation. The iron-reducing bacterial (IRB) populations' density and Fe^{2+} content in the paddy soil were monitored during the cultural cycle.

2. MATERIALS AND METHODS

2.1. Sampling Site

The site for experiments was located at Kamboinse in the central plateau of Burkina Faso (12°26'48"N and 1°33'45"W). Experiments were performed on the experimental site of the International Institute for Water and Environmental Engineering (2iE) from August to December 2011. Physical and chemical properties of the soil collected from Kamboinse, are presented in Table 1. The soil studied had clay-loam texture, a high content of sulfates and iron, had a low organic matter content and wasn't deficient in crop nutrients as N, K and Mg.

2.2. Microplots Experiments

Twelve microplots of 1.50 m × 1.20 m × 1.20 m in size (total internal volume: 1.31 m^3) were constructed in reinforced concrete to prevent cracks due to temperature changes following the alternation of the rainy and the dry seasons [19] . At the bottom of each microplot (1.45 m depth), perforated PVC pipes were installed and con- nected to an external tap through the concrete wall to sub-drain the soil (Figure 1). The design intended to improve circulation of water and oxygen within the rice root zone. Microplots were filled with the clay-loam soil described above, then flooded for 1 week. After the flooding period, two 15-day-old plants of FKR 19 rice (6 - 7 tons/ha of potential yield), developed by the Institute for Environment and Agricultural Research (INERA) from Oryza sativa rice, were transplanted at 20 cm hill intervals. Then, the soil was continuously flooded until rice maturity and harvest. Three replications and four modes of drainage were performed throughout the study: without drainage (control: T) and drainage for 7 days (P1), 14 days (P2) and 21 days (P3), respectively by regulating the drained water flow from the bottom tap (Figure 1). Chemical fertilizers (NPK and urea) were applied in the microplots according to the recommended doses of 300 kg/ha of N-P-K (14-23-14) and 100 - 150 kg/ha of urea [20] [21] .

2.3. Enumeration of Iron-Reducing Bacterial Populations

Iron-Reducing Bacteria (IRB) numbers were determined by the most-probable-number (MPN) method.

Medium for enumeration: The basal medium for enumeration was adapted from Hammann and Ottow [22] and consisted of the following components (per liter of distilled water): Glucose (20 g), Sodium acetate (5 g), $MgSO_4 \cdot 7H_2O$ (0.05 g), K_2HPO_4 (0.25 g), KH_2PO_4 (0.25 g), NaCl (0.05 g), $Na_2MoO_4 \cdot 2H_2O$ (0.005 g), $CaCO_3$ (5 g), Fe_2O_3 (1 g). The pH was adjusted to 7.2 by addition of NaOH before 9 ml were dispensed into 34 ml Hungate tubes. The tubes were then autoclaved at 121˚C for 15 min.

Soil dilution and enumeration conditions: Soil (0 - 15 cm) samples were collected before transplanting and near rice hills after transplanting with a small corer and immediately introduced into air-free 10 ml Veinoject tubes. Tubes were carried to the laboratory and kept at 4˚C until bacterial enumeration. Soil (5 g) from tubes was ground using a sterilized pestle and mortar and suspended in 45 ml sterile 0.5% (w/v) $MgCl_2$, $6H_2O$ solution. One milliliter of the mixed soil suspension was placed into 9 ml sterile 0.5% $MgCl_2 \cdot 6H_2O$ solution in Hungate tubes and diluted successively to 10^{-9} diluted suspension. One milliliter of the 10^{-1} to 10^{-9} fold diluted suspen-

Table 1. Physico-chemical characteristics of Kamboinse soil.

Gravel (%)	Sand (%)	Loam (%)	Clay (%)	MO (%)	Total C (%)	Total N (%)	C/N	K (ppm)	Fe_t (g/Kg)	SO_4 (ppm)	Mg (ppm)	pH_{H_2O}
34	34	10	22	1.945	1.128	0.105	11	2520	14.51	118.81	681.52	6.52

MO, Organic matter; Fe_t, total amount of Fe.

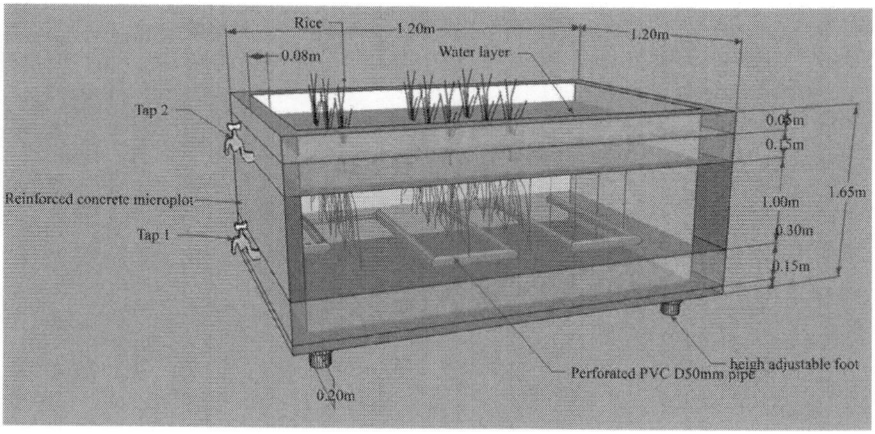

Figure 1. Microplots design for the experiments [19] .

sion was inoculated to 9 ml of the basal medium for enumeration in Hungate tubes. Three tubes per dilution were used. Tubes were incubated at 30°C in the dark for 5 days, then immediately analyzed for Fe^{2+} after addition of 0.2% ortho-phenantroline and 10% acetic acid according to Prade [23] . Positive tubes sustaining the reduction of Fe^{3+} (Fe_2O_3) into Fe^{2+} (reddish coloration) were revealed by measuring optical density at 510 nm using a spectrophotometer. The MPN number of IRB was determined using a MPN table for three tubes according to Hugues and Plantat [24] . Bacterial enumeration was performed before flooding when the soil was dried, on transplanting day (1 week after flooding) and during the rice growth stages until harvest near rice hills.

2.4. Determination of Ferrous Iron Content in Soil

From the soil sampled for bacterial enumeration and at the same periods during the rice cultural cycle, Vizier [25] method was used to measure the content of ferrous iron in the soil solution, using 0.2% ortho-phenantroline and 10% acetic acid reagents as described above.

2.5. Statistical Analysis

Data obtained were analyzed for IRB populations' development and activity, drainage mode, rice growth stage and rice yield variations using the Student's t-test and XLSTAT 7.5.2 statistical software. Mean parameters were compared according to the Newman Keuls' test at 5% probability level.

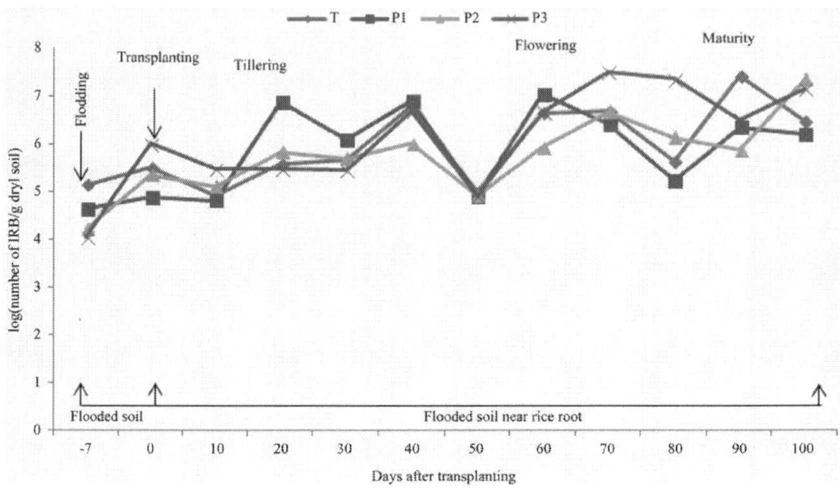

Figure 2. Densities of Iron-Reducing Bacteria in soil before flooding, at transplanting day, and in soil near rice roots during the cultural cycle of FKR 19 rice in microplots without drainage (T), and drained for 7 days (P1), 14 days (P2) and 21 days (P3), respectively (means of 3 replicates).

3. RESULTS AND DISCUSSION

3.1. Effect of Soil Moisture on IRB Populations Dynamic and Ferrous Iron Content in Soil near Rice Roots

Densities of IRB enumerated on soil before flooding, when the soil was dried and on soil near rice roots during the rice growth stages, are shown in Figure 2 and expressed as $\log_{10}$ (most probable population number g^{-1} dry soil). The number of IRB in the soil near rice roots increased after one week of flooding for all microplots (Figure 2). An increase in IRB population number of 124.63%, 40.89%, 52.84% and 98.92% was recorded in the soil of the control (T), P1, P2, and P3 drained microplots, respectively. Hammann and Ottow [22] reported that soon as a soil is flooded or submerged by stagnant water, the reductive processes start [26] [27] . During the process, oxygen is consumed by soil respiration (bacteria and fungi mainly) and a large amounts of mineral elements as Fe III and nutriments were released in the soil solution [28] . In these conditions Fe III was used as electron acceptor by IRB, for anaerobic respiration coupled to organic compound biodegradation [7] [8] [29] -[31] . Our findings are in agreement with the results of Ouattara [6] which reported that lactate-

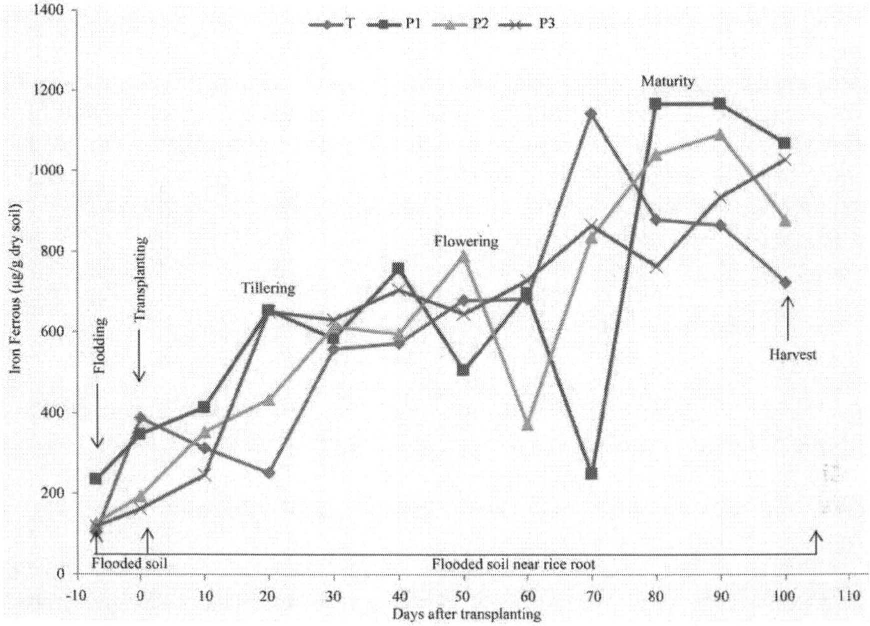

Figure 3. Evolution of the soil Fe^{2+} content during the cultural cycle of FKR 19 rice in microplots not drained (T), and drained for 7 days (P1), 14 days (P2) and 21 days (P3), respectively (means of 3 replicates).

utilizing IRB enumerated in Burkina Faso paddy soil (Kou Valley), ranged from 3.5×10^2 - 7.3×10^2 cells/g dry soil before flooding to 2×10^4 - 6×10^7 cells/g dry soil after flooding. As also underlined Jacq et al. [17] and Betremieux [32] , the growth of Iron-Reducing Bacteria is stimulated by flooding.

Fe^{2+} content in soil before flooding when the soil was dried and in soil near rice roots during the rice growth stages, are shown in Figure 3 and expressed as $\log_{10}$ (µg/g dry soil). The concentration of ferrous iron increased after one week of flooding in the soil near rice roots for all microplots (Figure 3). After flooding, a concentration increase of 73.07, 31.9, 34 and 26.53% was recorded in the soil of the control and in P1, P2, P3 drained microplots, respectively.

The concentration of ferrous iron increased after one week of flooding in the soil for all the microplots (Figure 3). This result is in agreement with the findings of Jacq et al. [17] and Betremieux [32] who reported the same increment of ferrous iron content in soil after the flooding of a Senegal paddy soil.

Iron toxicity only occurs in flooded soils and affects primarily the production of lowland rice [9] [11] . Cherif et al. [1] and Shahid et al. [11] underlined that the reducing conditions of waterlogged lowland soils boost iron toxicity through solubilization of almost all iron into its ferrous form (Fe^{2+}). After flooding, oxygen is depleted rapidly by the respiration of soil microorganisms [33] . With the depletion of oxygen, Fe^{3+} can act as electron acceptor for microbial respiration and sequentially become reduced. Within a few days after inundation or at a redox potential of <180 - 150 mV, the reduction of Fe^{3+} begins [7] [17] [34] .

3.2. Effects of Rice Plant on IRB Populations' Dynamic and Activity

Jacq et al. [17] , reported that the amount of ferrous iron in the soil solution of flooded soils are significantly higher in planted than in unplanted soil. In the present study, the number of IRB and ferrous iron content in soil near rice roots increased gradually with fluctuations from transplanting day to rice flowering and maturity stages in all the paddy microplots (Figure 2 & Figure 3). In most microplots, the highest densities of IRB were recorded from rice tillering and flowering to maturity stages, at which bacterial population number could reach 10^6 to 10^7 /g dry soil. The highest levels of ferrous iron in soil near rice roots were recorded also from rice tillering and flowering to maturity stages at which it could reach 10^3 to 10^4 µg/g dry soil. IRB population density and ferrous iron concentration in soil near rice roots appeared particularly high during the tillering and flowering to maturity stages of rice growth (Figure 2 & Figure 3). Our results are in agreement with those obtained by Jacq et al. [17] who recorded also a similar evolution pattern of IRB population in Senegal paddy soils during rice cultural cycle. However, these findings are not in agreement with the results of Ouattara [6] , who reported constant IRB populations in a rice paddy soil of Burkina Faso (Kou Valley) after 21 days of flooding. The high physiological activities of rice plant in tillering, flowering and maturity stages, may result in the production of more substrates available for bacterial growth [28] . These periods which correspond to the highest level of reduced soil condition may enhance the exudation of carbohydrates and other metabolites sustaining IRB population growth and ferrous iron highest content in soil [8] [17] .

3.3. Impact of Drainage on IRB Populations' Development and Activity

The variance of IRB numbers in soil near rice roots, in relation to drainage and sampling period during rice growth is presented in Table 2. The Newman Keuls' test revealed that the number of bacteria in the paddy soil near rice roots was significantly related to drainage ($p = 0.024$), sampling period ($p < 0.0001$) and combined both factors ($p < 0.001$, Table 2). Throughout the experiment, the average number of IRB population in the soil of microplots decreased significantly ($p = 0.024$) with drainage of water in P1 and P2 microplots, relatively to the control (T) and P3 ones (Table 3). The rate of decrease was 2.2% and 3.24% in P1 and P2 drained micro- plots relatively to the control, respectively. However, no significant difference was observed between IRB popu- lations in the soil of T and P3 microplots (Table 3).

The variance of soil Fe^{2+} content in soil in relation to drainage and sampling period during rice growth is presented in Table 2. It appeared that drainage alone had no significant effect on the soil near rice roots ferrous

Table 2. Variance of IRB number and ferrous iron content in soil near rice roots in relation to drainage and sampling period during the cultural cycle of FKR 19 rice.

Source of variation	df	log (IRB number/g dry soil)		Fe II (µg/g dry soil)	
		F	p	F	p
Drainage	3	3.292	0.024[*]	0.464	0.708[ns]
Period	11	22.153	<0.0001[**]	21.759	<0.0001[**]
Drainage x Period	33	2.249	0.001[**]	1.935	0.007[**]

df = degree of freedom; F = Fisher F; [*]significant $p < 0.05$; [**]significant $p < 0.01$; [ns]not significant $p < 0.05$.

Table 3. Effect of drainage on IRB population number in soil near rice roots during the cultural cycle of FKR 19 rice in microplots not drained (T), and drained for 7 days (P1), 14 days (P2) and 21 days (P3) (means of 3 replicates).

Drainage	log (IRB number/g dry soil)	Fe II (µg/g dry soil)
T	5.867[a]	597.989[a]
P1	5.630[b]	655.012[a]
P2	5.570[b]	625.588[a]
P3	5.957[a]	624.462[a]

Means with a same letter within a column are not significantly different according to Newman Keuls' test $p < 0.05$.

Means with a same letter within a column are not significantly different according to Newman Keuls' test $p < 0.05$.

iron content ($p = 0.708$). However, the sampling period in relation to rice growth stage affected significantly the soil ferrous iron content (p <

0.0001). The combined effects of drainage and sampling period also affected significantly ($p = 0.007$) the content of ferrous iron in soil near rice roots.

Ferrous iron content evolution in soil during the cultural cycle of rice appears similar for drained and not-drained microplots (Figure 3).

Most of the highly active Iron-Reducing Bacteria are anaerobe or facultative anaerobe. IRB are very efficient in anoxic conditions, and thus oxygenation reduces their growth [17] . However, the IRB population number in the soil of the P3 drained microplots increased slightly (1.01%) indicating that some among the IRB population can survive and grow in aerobic compartments in the presence of low pressure of oxygen where a surplus of oxygen is released by healthy roots [17] [23] .

Survival of IRB populations and ferrous iron production in drained microplots could be, also explained by heterogeneous distribution of oxygen through the drainage creating anoxic compartments. A lack of oxygenation of these microsites in drained microplots promotes IRB population survival [7] and ferrous iron production [3] . An efficient oxygenation of soil near rice roots by water drainage would significantly reduce anaerobic IRB population's activity.

3.4. Impact of Drainage on FKR 19 Rice Yield

Becker and Folkard [9] and Shahid et al. [11] reported that symptoms of iron toxicity in rice result from excessive uptake of ferrous ions by the roots and their acropetal translocation through xylem outflow to the aerial parts. According to these authors, the typical symptoms linked to this are the bronzing or yellowing processes of the leaves [9] [11] . In the present study, symptoms of rice iron toxicity weren't observed in all the microplots. One raison may be the FKR 19 rice variety is not susceptible to iron toxicity as reported some studies [19] . As also reported Keita [19] , even the oxygenation in the root zone was not enough to prevent the growth of Iron-Reducing Bacteria and the ferrous iron concentration, it did help in increasing the soil pH (from 5.7 to 7.3) which will limit the absorption of ferrous iron by the rice and thus, improve rice yield [11] [35] . Rice plants can develop also physiological avoidance and/or tolerance mechanisms [7] [36] [37] to survive under Fe-toxic condition such as: oxidation of ferrous iron at the root surface (Fe-oxidizing power) [38] , exclusion of Fe at the root surface (Fe-excluding power) [39] [40] , retention of Fe in the root tissue (Fe-retaining power) [39] [40] or leaf tissue tolerance to excess amounts of Fe [41] .

Chemical fertilizers (NPK and urea) application in microplots may also explain these results [11] . Many studies reported that amendment of paddy

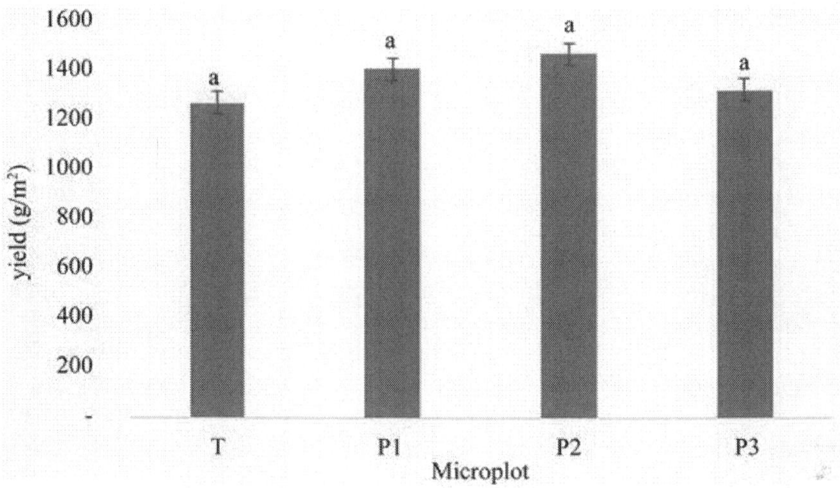

Figure 4. Yield of FKR 19 rice variety in microplots not drained (T), and drained for 7 days (P1), 14 days (P2) and 21 days (P3) (means of 3 replicates). Yields sharing the same letter are not significantly different according to Newman Keuls' test $p < 0.05$.

soil in mineral element may reduce symptoms of iron toxicity for rice [37] [42] [43] . In fact, iron toxicity is also defined as a multiple nutritive disorder further increased by mineral deficiencies through excesses of H_2S [13] [44] [45] . Deficiencies in minerals may enlarge the root permeability that enhances exudation, oxygen-consumption and iron-reduction, finally leading to severe iron toxicity [46] - [48] .

In the present study, the symptoms of iron toxicity for rice plant were not observed according IRRI standard evaluation systems for rice [49] in all the microplots and no significant difference ($p = 0.209$) was found for rice yield among treatments. However, drained microplots, in particular P2 microplot, showed the highest yield (Figure 4). These results are in agreement with those of Keita [19] and Ethan et al. [45] which reported that drainage in iron-toxic soils increased grain yield of lowland rice by the reduction of ferrous iron absorption of rice plant. Indeed, in oxygenated rice soil, the greatest fraction of iron (II) is present in precipitated forms (ferrous iron inactive forms) as siderite [7] [45] . Soil oxygenation by water drainage affected also iron (II) profiles by iron (II) oxidization in a fast chemical and microbial reaction to iron (III) in zones where oxygen is present [7] [50] .

4. CONCLUSION

The study revealed that flooding decreases Iron-Reducing Bacteria population's number in soil near rice roots. Despite the aeration of paddy soil by drainage, IRB population's number and activities appeared to be modulated mainly by the physiological activity of rice plant. Thus, ferrous iron content and IRB number were high in soil near rice roots during tillering, flowering and maturity stages. IRB activity did not exhibit a deleterious effect on rice yield due probably to the soil quality, drainage, chemical fertilizers application in microplots (NPK and urea) and/or to the rice variety (FKR 19). Although, no significant difference was observed for rice yield among drainage treatments, drained microplots showed the highest yields. Thus, an efficient oxygenation of soil by sub- drainage may be a key factor for ferrous iron toxicity control in paddy field.

ACKNOWLEDGEMENTS

The authors would like to express profound gratitude to International Institute for Water and Environmental Engineering (2iE), CNRST/IRSS, International Foundation for Science, PACER-UEMOA/RABIOTECH, FCN-WAAPP and CNS-FL/WAAPP, for financial and technical supports.

REFERENCES

1. Cherif, M., Audebert, A., Fofana, M. and Zouzou, M. (2009) Evaluation of Iron Toxicity on Lowland Irrigated Rice in West Africa. Tropicultura, 27, 88-92.
2. Engel, K., Asch, F. and Becker, M. (2012) Classification of Rice Genotypes Based on Their Mechanisms of Adaptation to Iron Toxicity. Journal of Plant Nutrition and Soil Science, 175, 871-881.
3. Ponnamperuma, F.N. (1972) The Chemistry of Submerged Soils. Advances in Agronomy, 24, 29-96.
4. Sahrawat, K.L. and Diatta, S. (1995) Nutrient Management and Season Affect Soil Iron Toxicity. Annual Report 1994. West Africa Rice Development Association, Bouake, Ivoiry Coast, 34-35.

5. WARDA (2002) Annual Report 2001-2002. Activity Highlights, ADRAO/WARDA., Bouake, Ivoiry Coast, 29-37.

6. Ouattara, A.S. (1992) Contribution to the Study of Iron-Reducing Bacteria and Sulfate in Paddy Soils of the Kou Valley (Burkina Faso). Ph.D Dissertation, Provence University, Aix-Marseille I.

7. Liesack, W., Shnell, S. and Revsbech, N.P. (2000) Microbiology of Flooded Rice Paddies. FEMS Microbiology Reviews, 24, 625-645.

8. Yi, W., Wang, B. and Qu, D. (2012) Diversity of Isolates Performing Fe(III) Reduction from Paddy Soil Fed by Different Organic Carbon Sources. African Journal of Biotechnology, 11, 4407-4417.

9. Becker, M. and Folkard, A. (2005) Iron Toxicity in Rice Conditions and Management Concepts. Journal of Plant Nutrition and Soil Science, 168, 558-573.

10. Mongon, J., Konnerup, D., Colmer, T.D. and Rerkasem, B. (2014) Responses of Rice to Fe2+ in Aerated and Stagnant Conditions: Growth, Root Porosity and Radial Oxygen Loss Barrier. Functional Plant Biology, 41, 922-929.

11. Shahid, M., Nayak, K.A., Shukla, A.K., Tripathi, R., Kumar, A., Raja, R. Panda, B.B., Meher, J., Bhattacharyya, P. and Dash, D. (2014) Mitigation of Iron Toxicity and Iron, Zinc, and Manganese Nutrition of Wetland Rice Cultivars (Oryza sativa L.) Grown in Iron-Toxic Soil. Clean Soil, Air, Water, 42, 1604-1609.

12. Paul, K., Diatta, M. and Millar, D. (Editors) (2010) The Effect of Iron Toxicity on the Yield and Yield Components of rice. Proceedings of Africa Rice Congress on Innovation and Partnerships to Realize Africa's Rice Potential, WARDA, IER, Bamako, 22-26 March 2010, 52-53. http://www.warda.cgiar.org/publications/ARC2010/ARC2010-Abstract.pdf

13. Fageria, N.K., Santos, A.B., Barbosa, F.M.P. and Guimar&atildees, C.M. (2008) Iron Toxicity in Lowland Rice. Journal of Plant Nutrition, 31, 1676-1697.

14. Kosaki, T. and Juo, A.S.R. (1986) Iron Toxicity in Inland Valleys: A Case Study from Nigeria. In: Juo, A.S.R. and Lowe, J.A., Eds., Wetlands and Rice in Sub-Saharan Africa Ibadan, IITA, Ibadan, 167-174.

15. Okusami, T.A. (1986) Properties of Some Hydromorphic Soils in West Africa. In: Juo, E.S.R. and Lowe, J.A., Eds., Wetlands and Rice in Sub-Saharan Africa Ibadan, IITA, Ibadan, 167-179.

16. Audebert, A. and Sahrawat, K.L. (2000) Mechanism for Iron Toxicity Tolerance in Lowland Rice. Journal of Plant Nutrition, 23, 1817-1885.

17. Jacq, V.A., Prade, K. and Ottow, J.G. (1991) Iron Sulphide Accumulation in the Rhizosphere of Wetland Rice (Oriza sativa L.) as the Result of Microbial Activities. In: Fyfe, W.S. Ed., Developments in Geochemistry, Vol. 6, Diversity of Environmental Biogeochemistry, Elsevier, Amsterdam, 453-468.

18. Audebert, A. (2006) Diagnosis of Risk and Approaches to Iron Toxicity Management in Lowland Rice Farming. In: Audebert, A., Narteh, L.T., Kiepe, P., Millar, D. and Beks, B., Eds., Iron Toxicity in Rice-Based Systems in West Africa, ADRAO, Cotonou, 6-17.

19. Keïta, A. (2015) Subsurface Drainage of Valley Bottom Irrigated Rice Schemes in Tropical Savannah: Case Studies of Tiefora and Moussodougou in Burkina Faso. PhD Thesis, Wageningen University, Delft.

20. INERA (2000) Descriptive Folders of Rice Cultivars. Institute of Environment and Agricultural Research, Ouagadougou. http://www.inera.bf/

21. Sokona, M.E.B., Boro, A., Hema, A. and Katiella, B. (2010) Diagnostic Study of the Rice Irrigation Scheme of Tiefora, Province of Comoe, Region of the Cascades. Field Report, 2iE, Ouagadougou.

22. Hammann, R. and Ottow, J.C.G. (1974) Reductive Dissolution of Fe2O3 by Saccharolytic Clostridia and Bacillus polymyxa under Anaerobic Conditions. Journal of Plant Nutrition and Soil Science, 137, 108-115.

23. Prade, K. (1987) Influence of Nutrient Supply to the Iron Poisoning of Paddy (O. Sativa L.) in the Basse Casamance Senegal. PhD Thesis, University of Hohenheim, Stuttgart. (In German)

24. Hugues, B. and Plantat, J.L. (1983) Calculation of the Acceptable Limits of the Most Probable Number When the Number of Inoculum Per Dilution Is Important. Chemosphere, 12, 1679-1684.

25. Vizier, J.F. (1969) Selection and Development of a Method for Iron Ferrous Dosage in Waterlogged Soils. ORSTOM Books, Pedology Serial, Montpellier.

26. Conrad, R. (2007) Microbial Ecology of Methanogens and Methanotrophs. Advances in Agronomy, 96, 1-63

27. Hori, T. Muller, A., Igarashi, Y., Conrad, R. and Friedrich, M.W. (2010) Identification of Iron-Reducing Microorganisms in Anoxic

Rice Paddy Soil by 13C-Acetate Probing. The ISME Journal, 4, 267-278.

28. Bongoua, D.A.J. (2009) Iron-Reducing Bacterial Communities Implications and Environmental Settings in the Functioning and Rice Fields Soils Quality (Thaïlande and Ivoiry Coast). PhD Thesis, Henri Poincare University, Nancy.

29. Berthelin, J. (1982) Microbial Processes in Hydromorphic Soils in Temperate Regions: Implications of Pedogenesis. Pedology Grand, 32, 313-328.

30. Bousserrhine, N., Gasser, U.G., Jeanroy, E. and Berthelin, J. (1999) Bacterial and Chemical Reductive Dissolution of Mn-, Co-, Cr-, and Al-Substituted Goethites. Geomicrobiology Journal, 16, 245-258.

31. Ehrlich, H.L. (2002) Geomicrobiology. 4th Edition, Marcel Dekker Publisher, New York.

32. Betremieux, R. (1951) Experimental Study of Iron and Manganese in the Soils. Annales Agronomiques, 10, 193-295.

33. Prade, K., Ottow, J.C.G., Jacq, V.A., Malouf, G. and Loyer, J.Y. (1990) Relationships between the Properties of Flooded Rice Soils and Iron Toxicity in Lower Casamance (Senegal). Studies, Review and Summary of Previous Work. ORSTOM Books, Pedology Serial, IRD, Montpellier, 453-474.

34. Patrick, J.W.H. and Reddy, C.N. (1978) Chemical Changes in Rice Soils. In: IRRI, Ed., Soils and Rice, The International Rice Research Institute, Manila, 361-379.

35. Dave, G. (1985) The Influence of pH on the Toxicity of Aluminum, Cadmium, and Iron to Eggs and Larvae of the Zebrafish. Brachydanio Rerio Ecotoxicology and Environmental Safety, 10, 253-267.

36. Nozoe, T., Agbisit, R., Fukuta, Y., Rodriguez, R. and Yanagihara, S. (2008) Characteristics of Iron Tolerant Rice Lines Developed at IRRI under Field Conditions. Japan Agricultural Research Quarterly, 42, 187-192.

37. Sahrawat, K.L. (2010) Reducing Iron Toxicity in Lowland Rice with Tolerant Genotypes and Plant Nutrition. Plant Stress, 4, 70-75.

38. Ando, T. (1983) Nature of Oxidizing Power of Rice Roots. Plant Soil, 72, 57-71.

39. Tadano, T. (1975) Devices of Rice Roots to Tolerate High Iron Concentration in Growth Media. Japan Agricultural Research Quarterly, 9, 34-39.

40. Tadano, T. (1976) Studies on the Methods to Prevent Iron Toxicity in Lowland Rice. Hokudai Nougakubu Houbun Kiyou (Mem. Fac. Agric. Hokkaido Univ.), 10, 22-68.

41. Bienfait, H.F. (1989) Prevention of Stress in Iron Metabolism of Plants. Acta Botanica Neerlandica, 38, 105-129.

42. Sahrawat, K.L. (2004) Iron Toxicity in Wetland Rice and the Role of Other Nutrients. Journal of Plant Nutrition, 27, 1471-1504.

43. Panda, B.B., Sharma, S., Mohapatra, P.K. and Da, A. (2012) Application of Excess Nitrogen, Phosphorus, and Potassium Fertilizers Leads to Lowering of Grain Iron Content in High-Yielding Tropical Rice. Communications in Soil Science and Plant Analysis, 43, 2590-2602.

44. Ottow, J.C.G., Prade, K., Bertenbreiter, W. and Jacq, V.A. (1993) Iron Toxicity Mechanisms of Flooded Rice (Oryza sativa L.) in Senegal and Indonesia. In: Raunet, M., Ed., Lowland and Rice-Growing, CIRAD-CA, Montpellier, 231-241.

45. 45. Ethan, S., Odunze, A.C., Abu, S.T. and Iwuafor, E.N.O. (2011) Effect of Water Management and Nitrogen Rates on Iron Concentration and Yield in Lowland Rice. Agriculture and Biology Journal of North America, 2, 622-629.

46. Benckiser, G., Ottow, J.C.G., Watanabe, I. and Santiago, S. (1984) The Mechanism of Excessive Iron-Uptake (Iron Toxicity) of Wetland Rice. Journal of Plant Nutrition, 7, 177-185.

47. Benckiser, G., Santiago, S., Neue, H.U., Watanabe, I. and Ottow, J.C.G. (1984) Effect of Fertilization on Exudation, Dehydrogenase Activity, Iron-Reducing Populations and Fe (II) Formation in the Rhizosphere of Rice (Oryza sativa L.) in Relation to Iron Toxicity. Plant Soil, 79, 305-316.

48. Prade, K., Ottow, J.C.G. and Jacq, V.A. (1988) Excessive Iron Uptake (Iron Toxicity) by Wetland Rice (O.su tiva L.) on an Acid Sulfate Soil in the Casamance/Senegal. In: Dost, H., Ed., Selected Papers of the Dakar Symposium on Acid Sulphate Soils, I.L.R.I., Wageningen, 150-162.

49. IRRI (2002) Standard Evaluation System For Rice. International Rice Research Institute, Manila.

50. Hauck, S., Benz, M., Brune, A. and Schink, B. (2001) Ferrous Iron Oxidation by Denitrifying Bacteria in Profundal Sediments of a Deep Lake (Lake Constance). FEMS Microbiology Ecology, 37, 127-134.

10

CHAPTER

Treated Agro-Industrial Wastewater Irrigation of Tomato Crop: Effects on Qualitative/Quantitative Characteristics of Production and Microbiological Properties of the Soil

Giuseppe Gatta[a,*], Angela Libutti [a], Anna Gagliardi [a], Luciano Beneduce[a], Lorenzo Brusetti [b], Luigimaria Borruso [b], Grazia Discigliоa, Emanuele Tarantino[a]

[a]*Department of Agricultural Food and Environmental Science, University of Foggia, Foggia, Italy*

[b]*Faculty of Science and Technology, University of Bolzano, Bolzano, Italy*

ABSTRACT

A comparative study was carried out to evaluate the effects of two water irrigation sources on the quality and microbiological safety of tomato plants and fruit, and on the microbiological soil properties: irrigation with groundwater (GW) and with treated agro-industrial wastewater (TW). In a field experiment in southern Italy (Apulia region), the physico-chemical characteristics of the irrigation waters and the fruit quality parameters were determined. *Escherichia coli*, fecal *Enterococci* and *Salmonella* spp. were also monitored in the irrigation waters, tomato plant and fruit, and root-zone soil. Bacteriological analysis for total heterotrophic counts (THCs) were determined for plant, fruit, and soil samples. The irrigation water source did not significantly affect yield quantitative traits. However, with GW, the marketable fruit yield was higher than with TW (~82 vs. ~79 Mg ha^{-1}, respectively). For both irrigation treatments, the most important qualitative parameters that characterize the processing tomato fruit (i.e., dry matter content, pH, soluble solid content, color parameters) were in agreement with reports in the literature. For the microbiological results, the mean levels of *E. coli* and fecal*Enterococci* were 4408 and 3804 CFU 100 ml^{-1}, respectively, for TW (above the Italian guidelines for TW re-use). For the tomato plant and fruit, no *E. coli* isolated in either, and fecal coliforms and THC were not influenced by the irrigation waters ($P > 0.05$). Total bacterial enumeration by quantitative PCR was lower in soil irrigated with GW, than TW (3.69 vs. 4.02, $\times 10^6$, respectively). Moreover, soil microbial community patterns substantially differed between the two water treatments. These data show that while fecal indicators are not affected, the community composition and dynamics of the whole bacterial population in soil is influenced by the different qualities of these waters used for irrigation.

Keywords: Agro-industrial wastewater irrigation; Water quality; Tomato yield; Fruit quality;Microbial soil community; Fecal indicator

1. INTRODUCTION

The re-use of wastewater in agriculture is gaining wider acceptance in many parts of the world. It represents an agronomic option that is increasingly being investigated and taken up in regions with water scarcity, growing urban populations, and rising demand for irrigation water (Meli et al., 2002 and FAO, 2011). Many irrigated areas around the world are experiencing water shortages due to several factors, such as climate change and surface and groundwater pollution. Water scarcity poses serious economic, social and even political concerns in all of its aspects. Under these circumstances, treated wastewater use can help to mitigate the damaging effects of local water deficits (FAO, 2010).

Treated wastewater not only offers an alternative water irrigation source, but also the opportunity to recycle plant nutrients (Chen et al., 2008). Its application might ensure the transfer of fertilizing elements, such as nitrogen (N), phosphorous (P), potassium (K^+), organic matter, and meso-nutrients and micro-nutrients, into agricultural soil (WCED, 1987). Hence, wastewater nutrients can contribute to crop growth, although there is a need for their periodic monitoring, to avoid any imbalance in the nutrient supplies, which might cause excessive vegetative growth, uneven plant and/or fruit maturity, and/or reduced qualitative/quantitative aspects of yields (Pedrero et al., 2010).

Treated wastewater can also be a source of pathogenic organisms and potentially hazardous chemical substances, such as enteric bacteria and viruses, salts, heavy metals, and surfactants. These might then accumulate in the soils, with unfavorable effects on crop quality and productivity, and on the ecological soil conditions (Siebe and Cifuentes, 1995 and Chen et al., 2008). One of the major concerns with wastewater re-use is the risk of the transfer of pathogenic microorganisms that represent a potential risk to human health if they enter the food chain (Al-Lahham et al., 2003, Salgot et al., 2006,Toze, 2006 and Palese et al., 2009). Indeed, many studies have shown that microbiological contamination can be a major issue for the re-use of treated agricultural wastewater (Rubino and Lonigro, 2008, Lopez et al., 2010, Patterson et al., 2011 and Vivaldi et al., 2013). To maximize the benefits and at the same time, to minimize the risks related to the use of treated wastewater, international policies and uniform legislative frameworks should be adopted.

In Italy, the agricultural use of reclaimed wastewater (municipal and agro-industrial) is regulated by Ministerial Decree no. 185/2003. With regard to microbiological contamination levels in particular, this Decree has defined some significantly lower threshold values (e.g., *Escherichia coli*, <10 CFU 100 ml^{-1} in 80% of the samples) than those included in international guidelines. These threshold values can be considered highly restrictive, because the risk of contamination has been reported to be low when contamination of irrigation water does not exceed 1000 CFU ml^{-1} (WHO, 2006 and Blumenthal et al., 2000).

Studies have been carried out in southern Italy relating to treated urban wastewater re-use for the irrigation of crops (Pollice et al., 2004, Lonigro et al., 2007 and Lopez et al., 2007). These have included wastewater with microbiological contamination levels higher than those required by the current legislation, and they have indicated the opportunity to increase the threshold values in the Italian guidelines. Therefore, there is the need for further studies to better define acceptable microbiological contamination levels of different sources of irrigation water as used on different crops. These should also take into account wastewater treatment, irrigation methods, and cultivation practices.

The majority of the studies conducted on wastewater applications in agriculture have focused mainly on reclaimed urban effluents. The aim of the present study was to determine the effects of secondary treated agro-industrial wastewater on tomato crop performance. In particular, the objectives of the study were: (i) to evaluate the effects of the wastewaters on qualitative and quantitative aspects of tomato crop production; (ii) to assess the impact of the wastewaters on the microbiological contamination of tomato fruit and the microbiological soil properties.

2. MATERIALS AND METHODS

2.1. Field characteristics and agronomic conditions

This field trial was carried out with the tomato (*Solanum lycopersicum* L.; formerly*Lycopersicon esculentum* Mill.) cultivar 'Manyla' (Semillas Fitò, Spain) during the growing season of 2012 (April to August). It took place in an agricultural area in the Foggia district (Stornarella: 41° 15′N, 15° 44′E; altitude, 154 m a.s.l.) of the Apulian region in southern Italy, on a site belonging to the Fiordelisi agricultural and food manufacturing company, which produces and processes vegetables. The tomato plants were grown under a net house structure, covered with an anti-hail net, in six identical 15 m × 30 m plots that were located near to the company wastewater treatment plant.

The experimental area is characterized by a Mediterranean climate, with long-term mean annual rainfall of 590 mm, which is mainly distributed from October to April (Caliandro et al., 2005). The mean monthly main climate parameters recorded during the trial are reported in Table 1. These were measured by a weather station near to the experimental area, and stored on a nearby data-logger (Campbell Scientific, USA). The mean maximum and mean minimum temperatures during the growing season were 34.5 °C and 8.5 °C, respectively, and the total rainfall was 108.4 mm, of which about 62% (67.0 mm) occurred in the first month of the growing season.

The trial was carried out in a clay loam soil (United States Department of Agriculture classification), with a field capacity (−0.03 MPa) of 30.5% dry weight (dw), a wilting point (−1.5 MPa) of 15.9% dw, and a bulk density of 1.41 Mg m^{-3}. The main characteristics of the soil layer of the experimental site (0–60 cm) are as follows: sand, 40.1%; loam 32.5%; clay 27.4%; organic matter 1.6%; Olsen P_2O_5, 80.1 mg kg^{-1}; Ac-extractable K_2O, 730 mg kg^{-1}; total N, 0.8‰ (Kjeldahl); mineral NO_3–N, 4.75 mg kg^{-1}; mineral NH_4–N, 7.50 mg kg^{-1}; pH 7.9; electrical conductivity, 0.49 dS m^{-1}.

The tomato seedlings were transplanted into the plots on April 12, 2012, in mulched paired rows (40 cm apart) spaced at 250 cm, with the plants at a distance of 30 cm apart along each single row. The final plant density was 2.7 plants m^{-2}. The plants were grown in a vertical setting, using nylon threads disposed between plants collar and iron wires arranged longitudinally in the direction of the plant rows, and fixed to the upper part of the nethouse, at 2.5 m from the ground.

During the cropping season, standard agronomic practices for tomato crops in the area were performed. The soil was subsoiling to a depth of 45 cm,

Table 1. Main climatic parameters recorded during the growing season of the tomato crops (2012).

Month	Climatic parameter[a]						
	T_{min} (°C)	T_{max} (°C)	RH_{min} (%)	RH_{max} (%)	P (mm)	Ws (m s^{-1})	E_v (mm)
April	8.5	20.1	51.7	95.6	67.0	2.30	86.9
May	11.6	25.0	36.6	82.8	28.0	2.42	137.5
June	17.9	33.0	27.3	71.1	0.0	2.72	197.9
July	20.8	34.4	30.6	77.1	8.4	2.43	195.3
August	20.2	34.5	29.4	81.4	5.0	2.10	176.5
Growing season	**15.8**	**29.4**	**35.1**	**81.6**	**108.4**	**2.40**	**794.1**

[a] T_{min}, T_{max}, monthly minimum, maximum air temperature; RH_{min}, RH_{max}, monthly minimum, maximum relative air humidity; P, total precipitation; Ws, monthly mean wind speed; E_v, total "class A" pan evaporation.

and before transplanting, its surface was milled. Pre-transplanting fertilization was applied to the soil by distributing 35 kg ha^{-1} N and 70 kg ha^{-1} P_2O_5. Throughout the crop cycle, 75 kg ha^{-1} N and 100 kg ha^{-1} P_2O_5 were added through fertirrigation. Pest and weed control were performed according to local management practices.

The tomato fruit were hand harvested at full stage maturity. Four harvestings were performed from June to August, on the days after transplanting of: 82 (HD_1), 96 (HD_2), 110 (HD_3) and 124 (HD_4).

2.2. Treatments and experimental design

Two experimental irrigation treatments were applied to the tomato plants: irrigation with groundwater (GW), and irrigation with treated agro-industrial wastewater (TW). The GW was from a water source that is commonly applied for crop irrigation in the experimental area. The TW used in this study was taken from the wastewater treatment plant that purifies all of the wastewater produced by the company during their industrial processing of vegetables (i.e., tomatoes, egg plants, courgettes, peppers). It is an activated sludge wastewater treatment plant that produces an annual volume of effluent of approximately of 46,500 m^3. The incoming wastewaters undergo a preliminary treatment through a 6-mm sieve screen, to separate out the coarse organic waste. The effluent water then goes into an equalization tank, for the secondary biological treatment. At the end of this phase, the wastewater is clarified in a secondary settler, and the sludge is separated out. For the present study, part of the treated wastewater not subjected to chlorine treatment was directed into the experimental area through a 100-mm diameter PVC pipe, and stored in a 3000-l tank; subsequently, it was used for the tomato irrigation.

The experiment was laid out in a randomized complete block design with the two irrigation treatments each replicated three times (Fig. 1). A drip irrigation system was used for the crop irrigation. This comprised a single pipe, with drippers at a 2 l h^{-1} flow rate spaced every 40 cm, and it was arranged in the middle of each paired row. Except for the first irrigation that was designed for the rooting and establishment of the plants, the following irrigations were performed with each water treatment every time the available soil moisture was depleted to the threshold value of 40% (Allen et al., 1998). This irrigation scheduling took into account continuous measurements of volumetric soil water content changes at the effective rooting depth (soil layer depths: 0–10, 10–20, 20–30, 30–40, 40–50 cm), using frequency domain reflectometry probes (EasyAG, Sentek Sensor Technologies), which were installed in each plot prior to the crop transplanting. At each irrigation, the soil water content of each plot was increased to field capacity. The amount of irrigation water applied to the tomato crop during the whole crop cycle was 4957 m^3 ha^{-1}, with the water volume at each irrigation varying from 100 m^3 ha^{-1} to 300 m^3 ha^{-1}, depending on the growth stage of the crop.

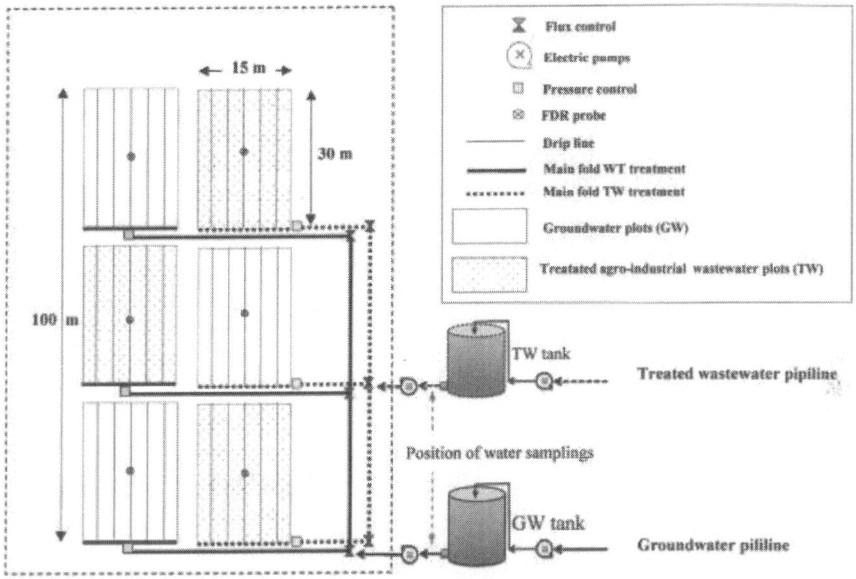

Figure 1. Layout of the experimental field and the irrigation systems.

2.3. Water, soil, plant and fruit sampling

GW and TW samples were collected at monthly intervals throughout the crop irrigation period, to characterize the physico-chemical and microbiological properties of these irrigation waters. Three samples of the GW and TW irrigation sources were collected (Fig. 1) in sterile 1000-ml glass bottles, and transported to the laboratory in refrigerated bags. The samples collected were kept in a refrigerator at +4 °C, and examined within 24 h of their collection.

Soil samples were collected in triplicate from the GW and TW plots, six times throughout the cropping season. All of the soil samples were taken from a 30 cm layer in each plot, from under the drippers, and they were air-dried, crushed, and passed through a 2 mm sieve before the chemical analysis. Tomato plant samples were collected at the same time, in triplicates for each experimental treatment.

The fruit sampling was performed at each of the four harvesting dates (about two days after irrigation crop) in an experimental plot of 20 m^2, by picking all of the mature fruit. The plants and harvested fruit were transported immediately to the laboratory for the chemical and microbiological analyses.

2.4. Water, soil and fruit chemical analysis

2.4.1. Water chemical analysis

The irrigation water samples were analyzed in triplicate, according to the Italian standard methods (APAT IRSA-CNR, 2003), which refer to the common international methods (APHA-AWWA-EF, 2005). The analysis included the physico-chemical parameters of pH, electrical conductivity (EC_w; dS m^{-1}), total suspended solids (TSS; mg l^{-1}), biological oxygen demand over 5 days (BOD_5; mg l^{-1}), chemical oxygen demand (COD; mg l^{-1}), ammonium-nitrogen (NH_4–N; mg l^{-1}), nitrate-nitrogen (NO_3–N; mg l^{-1}), nitrite-nitrogen (NO_2–N; mg l^{-1}), phosphorus (PO_4-P; mg l^{-1}), sodium (Na^+; mg l^{-1}), calcium (Ca^{2+}; mg l^{-1}), magnesium (Mg^{2+}; mg l^{-1}), potassium (K^+; mg l^{-1}), carbonates (CO_3^{2-}; mg l^{-1}), bicarbonates (HCO_3^-; mg l^{-1}), sulphate (SO_4^-; mg l^{-1}), sodium adsorption ratio (SAR) and water hardness (mg l^{-1} $CaCO_3$).

The pH was measured using a GLP 22+ pH & Ion Meter (CRISON INSTRUMENTS, Spain) and the electrical conductivity with a GLP 31+ EC-Meter (CRISON INSTRUMENTS, Spain). The Na^+, Ca^{2+}, Mg^{2+}, and K^+ levels were determined by ion-exchange chromatography (Dionex ICS-1100, Dionex Corporation, Sunnyvale, CA, USA). The TSS was determined after filtration of the water samples through 0.45-μm-pore-size (47-mm-diameter) nitrocellulose membranes (Whatman, Maidstone UK), using a vacuum system. The SAR was calculated using the formula (with concentrations in meq l^{-1}) (Richards, 1954): $SAR = (Na^+)/[(Ca^{2+} + Mg^{2+})/2]^{1/2}$.

2.4.2. Soil chemical analysis

The soil electrical conductivity and pH were measured on 1:2 (w/v) and 1:2.5 (w/v) aqueous soil extracts, respectively. The available phosphorus was determined using the sodium bicarbonate method (Olsen et al., 1954), and the total organic carbon (TOC) was determined by oxidation with potassium dichromate titration of $FeSO_4$, according toWalkley and Black (1934). The soluble NO_3–N and NH_4–N were determined according to Keeney and Nelson (1982).

All of the soil chemical parameters were used together with the soil microbiological characteristics for multivariate analysis, to determine the effects of the two soil treatments (i.e., GW, TW) on the dynamics of the bacterial communities.

2.4.3. Yield and fruit qualitative analysis

During the harvest, the marketable and discarded fruit were counted and weighted, to estimate the different components of the tomato yield: total yield (TY; Mg ha^{-1}), marketable yield (MY; Mg ha^{-1}), marketable fruit per plant (MYP; kg plant^{-1}), nonmarketable yield (NMY; Mg ha^{-1}), and nonmarketable fruit per plant (NMYP; kg plant^{-1}). On a sample of 10 marketable fruit from each plot, the following parameters were measured: mean diameter (equatorial and longitudinal diameter) (D; cm), soluble solids content of the flesh (SSC; °Brix), titratable acidity (TA; g citric acid 100 ml^{-1} fresh juice) (AOAC, 1995), dry matter content (DM; % fruit fresh matter) (AOAC 1990), a^*/b^* ratio (CI) (Jiménez-Cuesta et al., 1981), and Ca^{2+}, Na^+, Mg^{2+}, K^+ and nitrate NO_3^- content.

The color parameters were measured using a CM-700d spectrophotometer (Minolta Camera Co. Ltd., Osaka, Japan), as the CIELAB coordinates (i.e., L^*, a^*, b^*) on four randomly selected areas of the fruit surface. Here, only the a^*/b^* ratio is reported, which represents an index that describes well the color changes of tomato fruit (Francis and Clydesdale, 1975 and Favati et al., 2009).

The anion and cation contents were determined by ion-exchange chromatography (Dionex ICS-1100, Dionex Corporation, Sunnyvale, CA, USA). The anions were extracted from 0.5 g dried and ground samples, with 50 ml 3.5 mmol l^{-1} Na_2CO_3 and 1.0 mmol l^{-1} Na_2HCO_3, and they were measured using an IonPac AG14 precolumn and an IonPac AS14 separation column. The data are expressed as mg 100 g^{-1} fresh weight (fw). For the cations, 1.0 g dried and ground samples was use for the ash in a muffle furnace at 550 °C, and then digested in 20 ml 1.0 mol l^{-1} HCl in boiling water (99.5 ± 0.5 °C), for 30 min. The resulting solution was filtered, diluted, and analyzed using an IonPac CG12A guard column and an IonPac CS12A analytical column. The data are expressed as mg 100 g^{-1} fw (Renna et al., 2013).

2.5. Microbiological analysis

The GW and TW samples were analysed *E. coli* and fecal *Enterococci* enumeration, by the membrane filtration method. Triplicate aliquots of 100, 10, 1.0 and 0.1 ml of each water sample were filtered through 0.45-μm-pore-sized (47-mm-diameter) nitrocellulose membranes (Whatman, Maidstone UK). For *E. coli* enumeration, the membranes were placed onto tryptone bile agar with X-glucuronide (TBX agar; Oxoid, Basingstoke, UK) and incubated at 37 °C for 24 h. For fecal *Enterococci* enumeration, the membranes were placed onto Slanetz & Bartley Agar (Oxoid, UK), and incubated at 37 °C for 48 h. The same water samples were also analyzed for *Salmonella* spp., with their detection performed according to procedure UNI EN ISO 19250:2013.

The bacteriological analysis of the soil, plant and fruit samples included determination of*E. coli*, fecal coliforms, and total heterotrophic counts (THCs). These analyses were conducted by the spread plate method, as follows: 25.0 g of each sample was weighed and added to 225.0 ml buffered peptone water, homogenized in a stomacher for 180 s, and stored at room temperature for 30 min to allow bacterial cell recovery. Then serial 10-fold dilutions in buffered peptone water were spread onto plates containing TBX for*E. coli*, C-EC agar (Biolife) for fecal coliforms, and tryptic soy agar for THC. The plates were incubated under different incubation temperatures (and times): 37 °C for *E. coli*(24 h), 44 °C for fecal coliforms (48 h), and 22 °C for THC (72 h).

2.5.1. Quantification of total soil eubacteria by quantitative PCR

DNA extractions from triplicate soil samples irrigated with either GW or TW were performed with Powersoil DNA isolation kit (MoBio, Ca USA), following the manufacturer protocol. The DNA was eluted in 100 µl elution buffer, and visualized on ethidium bromide stained 1% agarose gels after electrophoresis, to assess the yield and quality of the extracted DNA. The DNA was quantified with an Eon microplate spectrophotometer (Biotek, Winooski, VE, USA) before further analyses.

Quantitative-PCR (q-PCR) analysis with universal primers targeting the bacteria was performed to determine the total eubacteria population in the soil. The primers and probes used were designed in previous studies (Nadkarni et al., 2002). Amplification and detection were performed using an AB 7300 real-time PCR system (Applied Biosystems, Foster city CA, USA), with a final reaction volume of 25.0 µl, which contained 100 nM of each primer, 150 nM probe, and 2× Taqman Fast Advanced master mix (Applied Biosystems). The cycling program was: 40 cycles of 15 s at 95 °C and 60 s at 60 °C. Conversion of the 16S rRNA gene copy numbers to cell numbers was carried out considering that bacteria have an average of four copies per cell of the 16S rRNA gene.

2.5.2. Soil microbial community analysis by automated ribosomal intergenic spacer analysis

To compare the soil microbial community compositions under the two different treatments (soil irrigated with GW and TW), automated ribosomal intergenic spacer analysis (ARISA) was applied. Internal transcribed spacer (ITS) regions of the soil microbial DNA were amplified using the primers ITSF (5′-GTCGTAACAAGGTAGCCGTA-3′) and ITSReub (5′-GCCAAGGCATCCACC-3′), labelled with 6-FAM, according to the chemical and thermal amplification protocol ofCardinale et al. (2004). The three replicates of six PCR samples from the soils (T_0, T_1, T_2, T_3, T_4, T_5) were sent to STAB Vida Lda (Caparica, Portugal) for the capillary electrophoresis. Peak Scanner Software 1.0 (Applied Biosystems) was used to analyze the fragment data. The T_0 and T_1 soil samples were taken 1 week before transplanting and 15 days after transplanting (just before starting the GW and TW irrigation), respectively, and T_2, T_3, T_4 and T_5 were collected at intervals of about one month (42, 69, 96 and 124 days from transplanting, respectively). To obtain the output matrix, each fragment size detected by ARISA was converted to the nearest integer value, and these were subsequently aligned according to their peaks, against the rounded sizes of the fragments, using the Microsoft-Excel macro Treeflap (Rees et al., 2004). The matrix was normalized and root-square transformed for the statistical analysis.

2.6. Statistical analysis

The measured data from each of the continuum variables relating to the qualitative/quantitative traits of the tomato fruit were processed statistically using analysis of variance (ANOVA). When significant effects were detected ($P \leq 0.05$), mean multiple comparisons were performed according to Tukey's tests. With reference to the analyzed qualitative parameters, the Bartlett test confirmed the homogeneity of the variance among the harvest data, so a combined statistical analysis was performed later. Furthermore, the variables related to the qualitative parameters of the tomato fruit were jointly considered in a multivariate approach, and statistically processed for canonical variates analysis (CVA), with the two experimental factors (water irrigation treatment and harvest data) as the discriminating sources (Cooley and Lohnes, 1971, Sadocchi, 1981,Gittins, 1985 and Podani, 2007). Before performing the CVA, the values of each variable were correctly standardized. The first two canonical variates accounted for the larger part of the data variability, and these were considered for further data interpretation. The CVA is represented graphically in a biplot, which considers both the canonical standardized scores (corresponding to each multivariate experimental datum) and the vectors (originating from the centroids and recording the canonical standardized coefficients). Finally, the ARISA data matrix and the standardized soil parameters were analyzed using canonical correspondence analysis (CCA). The ANOVA and CVA were performed using the JMP software package, version 8.1 (SAS Institute Inc., Cary, NC, USA). The CCA was performed using the PAST software in its default settings (Hammer et al., 2001). All of the graphical representations were carried out using the SigmaPlot software (Systat Software, Chicago).

3. RESULTS AND DISCUSSION

3.1. Irrigation water properties

Table 2 shows the means of the physico-chemical characteristics of GW and TW measured during the experimental trial. These analyses show some specific differences in their compositions. TW was characterized by higher N (especially as NH_4-N), P, Mg^{2+}, and K^+ than GW, which represent important nutrients for improving plant growth, soil fertility and crop yield. TW also showed higher organic matter content than GW (as indicated by BOD_5 and COD; Table 2). These higher nutrient levels in TW compared to GW indicate that this TW can provide an important source of plant nutrients for the soil, and can contribute to crop growth.

Table 2. Means of the main physico-chemical parameters measured during the experimental period for the groundwater (GW) and the treated agro-industrial wastewater (TW) used for the tomato irrigation.

Water parameter	Irrigation treatment		Significance
	GW	TW	
Ph	7.63 ± 0.10	7.76 ± 0.09	ns
EC (dS m^{-1})	0.69 ± 0.05	2.18 ± 0.12	*
TSS (mg l^{-1})	3.26 ± 0.34	16.21 ± 2.24	*
NH_4-N (mg l^{-1})	0.04 ± 0.00	0.39 ± 0.10	*
NO_3-N (mg l^{-1})	29.06 ± 1.67	1.20 ± 0.23	*
NO_2-N (mg l^{-1})	0.02 ± 0.01	0.07 ± 0.02	*
PO_4-P (mg l^{-1})	0.10 ± 0.01	0.29 ± 0.02	*
BOD_5 (mg l^{-1})	9.33 ± 1.03	21.58 ± 1.62	*
COD (mg l^{-1})	18.44 ± 1.62	39.73 ± 2.78	*
Na^+ (mg l^{-1})	33.53 ± 0.54	219.85 ± 6.05	*
Ca^{2+} (mg l^{-1})	52.82 ± 3.23	85.27 ± 1.24	*
Mg^{2+} (mg l^{-1})	8.90 ± 0.20	10.25 ± 0.12	*
K^+ (mg l^{-1})	9.35 ± 0.16	41.17 ± 1.96	*
CO_3^{2-} (mg l^{-1})	171.50 ± 5.32	193.67 ± 11.89	ns
HCO_3^- (mg l^{-1})	257.89 ± 2.85	254.23 ± 21.57	ns
SO_4^- (mg l^{-1})	30.17 ± 1.30	31.84 ± 0.85	ns
SAR	1.13 ± 0.03	5.99 ± 0.18	*
Hardness (mg l^{-1} $CaCO_3$)	168.57 ± 7.79	255.20 ± 3.54	*

Data are means ± standard errors for each analyzed trait, determined on 15 samples for each irrigation water treatment (1 sample per water treatment ×3 replicates ×5 sampling dates).

* Statistically significant at $P \leq 0.05$; ns, not significant.

For abbreviations, see main text.

Table 3. Influence of the groundwater (GW) and the treated agro-industrial wastewater (TW) used for the tomato irrigation on some of the quantitative traits of the tomato fruit.

Treatment	Quantitative trait[a]				
	Total yield	Marketable yield		Nonmarketable yield	
	(Mg ha^{-1})	Total (Mg ha^{-1})	Per plant (kg plant^{-1})	Total (Mg ha^{-1})	Per plant (kg plant^{-1})
GW	87.54 ± 10.37[a]	82.88 ± 9.47[a]	3.06 ± 0.29[a]	4.66 ± 0.89[a]	0.17 ± 0.05[a]
TW	85.32 ± 5.01[a]	79.05 ± 4.76[a]	2.93 ± 0.15[a]	6.26 ± 0.61[a]	0.23 ± 0.03[a]
Significance	ns	ns	ns	ns	ns

Data are means ± standard error, as measured from 162 plants (54 plants per plot ×3 replicates).
ns, not significant.
[a] Means followed by the same letters in each column are not significantly different ($P \leq 0.05$; Tukey tests).

However, of note, the NO_3–N of GW was significantly higher (29.06 mg l^{-1}) than for TW (1.20 mg l^{-1}) (Table 2). This elevated NO_3–N in GW appears to be due to an important nitrate contamination problem of the aquifer in the study area, where the intensive agricultural activity has led to the common and diffuse practice of extensive nitrogen fertilizer application to the various crops. This elevated NO_3–N content in GW represents an important source of nutrient for the plants, but generally it is not taken into account by the farmers when applying fertilizers. The resulting nitrogen surplus in the soil is then particularly exposed to the risk of leaching, thus increasing the environmental problem of nitrate pollution.

TW also showed higher Na^{+}, Ca^{2+}, Mg^{2+}, SAR, EC, and TSS than GW (Table 2). If the SAR is related to the EC (Ayers and Westcot, 1985) of TW, it appears that there is no limit to its agricultural application, and there would be no reduction in its rate of infiltration into the soil. Both GW and TW were alkaline, with a higher pH for GW than TW (although not significantly higher). The other parameters analyzed showed similar levels in GW and TW, and in general, the values for these main physico-chemical water properties met the Italian standards for wastewater re-use (Ministerial Decree no. 185/2003).

3.2. Effects of irrigation water on qualitative/quantitative traits of tomato yield

Table 3 gives the ANOVA data with reference to the influence of the water irrigation treatments on the productive traits of the tomato crop. These traits are related to the combined harvest dates (i.e., the cumulative yields).

For GW, MY was higher (82.88 Mg ha^{-1}), although not significantly so, than for TW (79.05 Mg ha^{-1}). This appears to be mainly due to the higher NMY for TW (6.26 Mg ha^{-1}) with respect to GW (4.66 Mg ha^{-1}). The MY recorded at the end of the tomato crop cycle is roughly in agreement with results obtained in other experimental trials carried out in other Italian regions (Aiello et al., 2007). The MY obtained from a tomato crop (variety 'Missouri') irrigated with fresh water by Aiello et al. (2007) was higher than that irrigated with wastewater. On the contrary, in the same study, for a different tomato variety (i.e., 'Incas'), the use of urban wastewater irrigation produced an increase in MY compared to the same tomato variety irrigated with fresh water (Aiello et al., 2007). Under the experimental conditions of the present study, this cultivar 'Manyla' tomato crop showed a lower yield than for this other study. This might be due to differences in the genetic constitution of the cultivars used, or to the type of treated wastewater applied and the pedo-environmental conditions of the cultivation area.

Table 4. Effects of water irrigation treatment, as the groundwater (GW) and the treated agro-industrial wastewater (TW), and harvest collection data (HD_{1-4}), and their factorial combination, on some of the qualitative traits of the tomato fruit.

Parameter	Qualitative traits[a]										
	DM	D	CI	SSC	TA	pH	Anion and cation content (mg 100 g^{-1} fresh matter)				
	(% FM)	(cm)	(a^*/b^* ratio)	(°Brix)	(g 100 ml^{-1})		Ca^{2+}	Mg^{2+}	K^+	Na^+	NO_3^-
Water irrigation treatment											
GW	7.44 ± 0.11^a	4.22 ± 0.32^a	1.03 ± 0.03^a	5.73 ± 0.12^a	0.31 ± 0.02^a	4.54 ± 0.04^a	8.63 ± 0.54^a	7.25 ± 0.63^a	225.3 ± 9.5^a	9.15 ± 0.94^b	1.32 ± 0.10^a
TW	7.52 ± 0.21^a	4.44 ± 0.28^a	0.97 ± 0.04^a	5.53 ± 0.12^a	0.30 ± 0.03^a	4.35 ± 0.04^b	9.33 ± 0.62^a	7.07 ± 0.33^a	225.4 ± 6.0^a	11.05 ± 0.92^a	0.92 ± 0.06^b
Significance	ns	ns	ns	ns	ns	*	ns	ns	ns	*	*
Harvest data											
HD_1	7.19 ± 0.11^a	5.04 ± 0.10^a	0.93 ± 0.06^a	5.40 ± 0.05^a	0.40 ± 0.01^a	4.36 ± 0.02^c	10.43 ± 0.88^a	7.22 ± 0.66	213.1 ± 17.8^a	7.21 ± 0.51^c	1.04 ± 0.08^a
HD_2	7.60 ± 0.21^a	4.63 ± 0.07^b	1.01 ± 0.02^a	5.60 ± 0.13^a	0.19 ± 0.02^c	4.48 ± 0.03^b	9.31 ± 1.03^a	6.45 ± 1.07	218.6 ± 8.8^a	7.93 ± 0.73^c	1.18 ± 0.14^a
HD_3	7.24 ± 0.31^a	3.82 ± 0.08^d	1.07 ± 0.04^a	5.85 ± 0.14^a	0.32 ± 0.01^b	4.43 ± 0.02^{bc}	8.45 ± 0.61^a	7.40 ± 0.47	228.4 ± 4.6^a	10.82 ± 0.71^b	1.12 ± 0.14^a
HD_4	7.98 ± 0.21^a	4.21 ± 0.06^c	1.17 ± 0.05^a	5.68 ± 0.27^a	0.30 ± 0.02^b	4.70 ± 0.03^a	7.73 ± 0.32^a	7.88 ± 0.53	241.5 ± 6.7^a	14.46 ± 0.66^a	1.14 ± 0.20^a
Significance	ns	***	ns	ns	***	**	ns	ns	ns	***	ns
Water irrigation treatment × Harvest data											
Significance	ns	ns	ns	ns	ns	ns	ns	ns	ns	ns	ns

Data are means ± standard error, as measured from 30 marketable fruits (10 fruit per plot, ×3 replicates).

* $P \leq 0.05$.

** $P \leq 0.01$.

*** $P \leq 0.001$; ns, not significant.

[a] Means followed by the same letters in each column are not significantly different ($P \leq 0.05$; Tukey tests).

For abbreviations, see main text.

Table 4 shows the means for the qualitative traits of the tomato yield with respect to the two experimental factors, the water irrigation treatments and the harvest data. The irrigation treatments and the harvest data did not show any significant effects in their interactions. The different water irrigation treatments significantly ($P \leq 0.05$) affected the pH of the tomato fruit, with a higher pH for GW (4.54) compared to TW (4.35). This pH parameter is very important, because it can strongly influence the effectiveness of thermal processes carried out on tomato fruit during their industrial transformation (Garcia and Barret, 2006). The pHs in the present study are comparable to those of another study related to the influence of irrigation and organic fertilization on fruit quality for tomato (Madrid et al., 2009); this other study showed a pH from 4.32 to 4.56, with the higher value in the fertilizer treatments (compared to the nontreated control).

The crop DM and SSC were not different between the GW and TW irrigation treatments (7.52% vs. 7.44%; 5.73 vs. 5.53 °Brix; respectively; Table 4). These data are in agreement with Turhan and Seniz (2009), who reported DM ranging from 4% to 7%, and SSC ranging from 3.3 °Brix to 5.5 °Brix for 33 genotypes of tomatoes cultivated in the Mediterranean region. However, our data here are lower than in another study (Sgherri et al., 2008), where for the Cherry tomato 'Dulcito RZ' grown in greenhouses, DM and SSC were 10% to 12% and 9 °Brix to 10 °Brix, respectively. However, in studies on the effects of irrigation on productivity and fruit quality of tomatoes produced under different water regimes, Patanè et al. (2011) and Favati et al. (2009) reported lower DM and SSC than in the present study. These differences will mainly be due to the different genotypes used (Sgherri et al., 2007 and Sgherri et al., 2008), and the environmental drought (Mahajan and Singh, 2006 and Soraya et al., 2001) and climate conditions (e.g., temperature, CO_2 concentration, light conditions). High DM and SSC might have important positive implications for the tomato canning and processing industry (Richardson and Hobson, 2006 and Favati et al., 2009), as it is well known that tomatoes with high SSC improve the processing efficiency, as less energy is needed to evaporate the water from the tomatoes when producing paste, concentrated juice, and dried or semi-dried tomatoes.

For the D, CI and TA parameters (Table 4), there were no significant differences between the TW and GW treatments, and these mean data are in agreement with other studies for similar tomato genotypes (Madrid et al., 2009 and Mahajan and Singh, 2006).

Among the mineral components of the tomato fruit (i.e., Ca^{2+}, Mg^{2+}, K^+, Na^+, NO_3^-), only the Na^+ and NO_3^- content showed differences with respect to water irrigation factors (Table 4). The Na^+ content of the fruit was higher for TW than GW (11.05 vs. 9.15 mg 100 g^{-1}, respectively); this is probably related to the higher content of Na^+ in TW compared to GW (see Table 2). Finally here, the NO_3^- content in the tomato fruit was higher for the GW treatment (1.32 mg 100 g^{-1}) than for TW (0.92 mg 100 g^{-1}), which is in agreement with the higher NO_3–N in GW, and it is also much lower than that defined in the European guidelines (Reg., 2006 and Reg., 2011). Except for the Ca^{2+} and Na^+contents, our data are in agreement with the mineral data for tomato fruit reported in other studies (Suárez et al., 2007 and Guil-Guerrero and Rebolloso-Fuentes, 2008).

For the harvest data experimental factor, only the means of the D, TA, pH and Na^+parameters were significantly different. Across all of the harvest data, D had a mean that ranged from 3.82 cm to 5.04 cm, and was significantly higher for HD_1 than for the rest of the harvest dates (HD_{2-4}). This can be explained by the position of the tomato fruit on the plant during the first harvest (HD_1). Indeed, our data are in agreement with the consolidated literature, in which it has been stated that the size and shape of a fruit can also vary in relation to the position of the fruit within the plant and the sequence of pollination among the flowers (Sawhney and Greyson, 1972 and Bohner and Bangerth, 1988). In particular, it has been shown that the first fruit on the first truss is generally larger in size than the rest, and that it can also be multilocular, which further supports the relationship between locule number and fruit size.

The TA fruit content was between 0.4 g 100 ml^{-1} and 0.19 g 100 ml^{-1} across the harvest dates, with a significant difference between the first two harvest dates (HD_1 vs. HD_2). The TA for HD_3 and HD_4 (0.32 vs. 0.30 g 100 ml^{-1}, respectively) were not different. The pHs were within the range of 4.36 to 4.60, at HD_1 and HD_4, respectively.

For the ionic components of the fruit, the Na^+ content gradually increased with the later harvest dates, from 7.21 mg 100 g^{-1} to 14.46 mg 100 g^{-1} for HD_1 and HD_4, respectively. As already indicate above, this increase in Na^+ is probably due to the high Na^+ in the water irrigation treatments (and particularly for TW) and will parallel the progressive Na^+accumulation in the soil.

3.2.1. Canonical analysis on the qualitative composition of the tomato fruit

The ANOVA results reported above show the effects of experimental factors on each qualitative variable separately. A multivariate approach analysis allowed us to integrate these data to evaluate which of the qualitative variables (considered simultaneously) contributes most to the group differences (experimental factors). According to the CVA, the eleven original qualitative variables related to the qualitative traits of the tomato fruit were reduced to two canonical variates that represent 76.2% of the total data variability: 53.1% for the first (CAN_1), and 23.1% for the second (CAN_2) (Table 5). On the basis of the large amount of overall variability explained by CAN_1, this can be considered as a 'multivariate qualitative index' of the tomato fruit.

Table 5. Standardized coefficients (scores) for the first two canonical variates (*CANj*), considering the qualitative properties of the tomato fruit. The corresponding percentages of accounted variation are also reported.

Original variable (*Vi*)	Standardized canonical coefficients	
	CAN_1	CAN_2
Mean diameter (D)	−1.167	−0.505
a^*/b^* ratio (CI)	−0.493	0.199
Soluble solids content of the flesh (SSC)	0.001	0.098
pH of the flesh	0.975	−1.051
Titratable acidity (TA)	−0.137	0.343
Dry matter (DM)	0.057	−0.074
Nitrate content (NO_3^-)	0.589	−0.226
Sodium content (Na^+)	−0.101	0.856
Potassium content (K^+)	−0.012	−0.113
Magnesium content (Mg^{2+})	−0.272	0.101
Calcium content (Ca^{2+})	0.187	0.070
Percentage explained variation	53.1	23.1
Percentage cumulative variation	53.1	76.2

To correctly interpret the relationships between canonical variates (*CANj*) and the original variables (*Vi*), it is important to recall that the *CANj* are linear combinations of the original variables and that the canonical coefficients maximize the discrimination among the experimental factors considered using canonical coefficients. The original variable *Vi*with the largest standardized canonical coefficient has, indeed, the strongest impact on the canonical variates *CANj*.

CAN_1 was positively affected by pH and NO_3^- content (scores, 0.975 and 0.589, respectively), and negatively affected by the D and a^*/b^* ratio (CI) parameters (scores, −1.167 and −0.493, respectively) (Table 5). The other *Vi* contributed to a limited extent to defining CAN_1 (low canonical scores). CAN_2, instead, was mainly defined by the Na^+content (score, 0.856), and to a different extent by pH and D of the fruit (scores, −1.051 and −0.505, respectively).

Coherent information can also be derived from the correlation matrix (Table 6) between the 'old' original variables (*Vi*) and the 'new' canonical variables (*CANj*). The higher the correlation coefficient between *Vi* and *CANj*, the stronger the influence that *Vi* has on*CANj*. These considerations can be easily seen using biplot graphs (Fig. 2) by considering the length and orientation of the 'vectors'. The biplots represent the effects of the discriminating experimental factors (irrigation water treatment and harvest date) on the qualitative characteristics of the tomato fruit.

Table 6. Correlation matrix (Pearson coefficients) between the quality parameters of the tomato fruit and the two canonical variables extracted (CAN_1 and CAN_2).

Original variables (*Vi*)	Canonical variables	
	CAN_1	CAN_2
Mean diameter (D)	−0.84**	−0.42*
a^*/b^* ratio (CI)	−0.39	−0.12
Soluble solids content of the flesh (SSC)	0.40	0.06
pH of the flesh	0.64**	−0.61*
Titratable acidity (TA)	−0.29	0.32
Dry matter (DM)	0.24	−0.24
Nitrate content (NO_3^-)	0.41*	−0.39
Sodium content (Na^+)	−0.36	0.55*
Potassium content (K^+)	−0.09	−0.11
Magnesium content (Mg^{2+})	−0.26	0.16
Calcium content (Ca^{2+})	0.30	0.15
CAN_1	1	0
CAN_2	0	1

* $P \leq 0.05$.
** $P \leq 0.001$.

Among the original qualitative variables considered in the CVA, those that affected CAN_1(pH, NO_3^-, D) and CAN_2 (Na^+, D, pH) were the same ones that better discriminated between the experimental factors (irrigation water treatment and harvest date), in agreement with the ANOVA results (see Table 4). In particular, Na^+ content allowed discrimination of the irrigation treatment with TW, while NO_3^- content and pH of the fruit allow identification of the irrigation treatment with GW. Thus, from the GW to TW water treatments, there is a marked increase in the Na^+ content and a corresponding decrease in both pH and NO_3^- content.

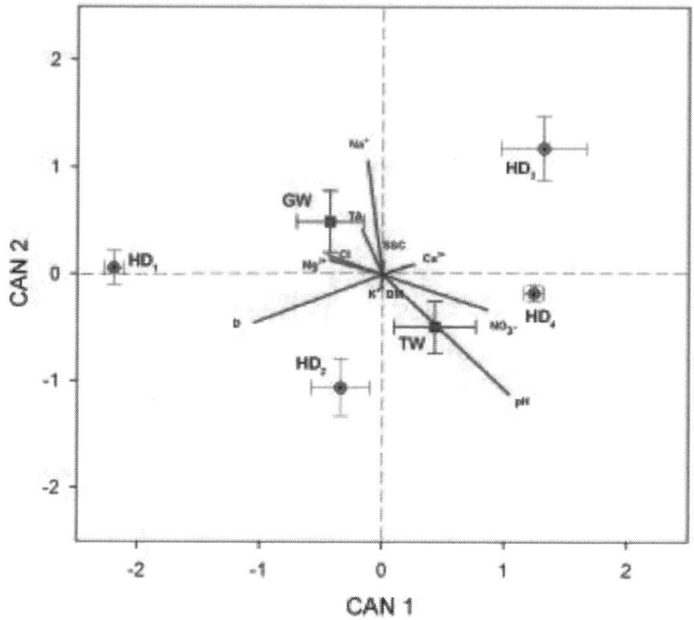

Figure 2. Canonical analysis of the qualitative parameters of the tomato fruit using the harvest data (HD_j) and the water irrigation treatments (GW, TW) as the discriminant variables. For abbreviations, see main text. Horizontal and vertical bars indicate standard errors of three replicates.

Considering the first canonical axis, the high D (low CAN_1) discriminates between the first two harvest dates (HD_1 vs. HD_2), while pH identifies the third harvest date (HD_3). Finally, the last harvest date (HD_4) is characterized by a lower D (i.e., its position is distant from D; Fig. 2). The other quality traits (DM, CI, SSC, Ca^{2+}, Mg^{2+}, K^+) showed low contributions to the discrimination of the experimental groups (irrigation water treatment and harvest date), considering the length of their 'vectors'.

In general, the results of the multivariate analysis show that pH and NO_3^- allow better differentiation between the two water irrigation treatments, while D and pH allow better differentiation between (most of) the harvest dates.

3.3. Effects on microbial indicators

With the aim of evaluating the safety aspects related to the use of these different qualities of irrigation water for tomato crop irrigation, several microbial analyses were conducted, as reported in Table 7. For TW, the mean *E. coli* and fecal *Enterococci* counts were 4400 and 3800 CFU 100 ml^{-1}, respectively. These are well above the Italian limits for treated wastewater re-use (Ministerial Decree no. 185/2003). These data are in agreement with the objective of the present study, as we intentionally provided an input of non-chlorinated treated wastewater to also be able to evaluate any effects on the soil and plant microbiological quality.

Table 7. Enumeration of the bacterial indicators in the water, soil, plant and fruit samples according to the groundwater (GW) and the treated agro-industrial wastewater (TW) irrigation.

Source	Irrigation Treatment		Significance
Bacterial indicator	GW	TW	
Water (CFU 100 ml^{-1})			
E. coli	7	4400	**
Fecal *coliforms*	9	3800	**
Soil (CFU g^{-1})			
E. coli	n.d.	n.d.	
Fecal *Enterococci*	1230	1360	ns
Total heterotrophic counts	3.69×10^6	4.02×10^6	ns
Soil q-PCR (cells g^{-1})			
Total eubacteria	4.86×10^7	16.2×10^7	*
Plant (CFU g^{-1})			
E. coli	n.d.	n.d.	
Fecal *Enterococci*	150	183	ns
Total heterotrophic counts	18,400	16,400	ns
Fruit (CFU g^{-1})			
E. coli	n.d.	n.d.	
Fecal *Enterococci*	173	237	ns
Total heterotrophic counts	5500	17,800	ns

* $P \leq 0.05$.
** $P \leq 0.01$.
ns, Not significant n.d., not detected.

GW was almost free of fecal indicators, with 7 CFU 100 ml^{-1} and 9 CFU 100 ml^{-1} for *E. coli* and fecal *Enterococci*, respectively. *Salmonella* spp. was not isolated in any of the water samples. In that case the microbiological parameters are in line with the above cited legislation limits.

For the irrigated soil, the level of fecal coliform counts was almost identical for GW treatment and TW treatment. The total heterotrophic plate counts (THC) showed a lower level in the soil irrigated with TW than GW (4.02×10^6 and 3.69×10^6, respectively), although this was not significant

($P > 0.05$). *Salmonella* spp. was not isolated in any of these water samples, and thus here the microbiological parameters are in line with the above-cited legislation limits.

When q-PCR was applied, the total bacterial level of TW (16.2×10^7 cells g^{-1}) was significantly higher than GW (4.86×10^7 cells g^{-1}) ($P \leq 0.05$), Moreover the q-PCR count estimated the total (viable and nonviable) bacterial populations in the soil as >1 log CFU g^{-1} than the viable cell count. The marked differences between the bacterial cultivation method (THC) and the molecular method (q-PCR) in assessing the total bacterial population arises from the higher sensitivity of the q-PCR, as a narrower range of the total bacterial population can be cultivable on synthetic media. *E. coli* was not isolated in any of the soil samples, independent of the water used for irrigation.

The data related to fecal coliforms and THC in the soil are in agreement with Benami et al. (2013), who assessed soil irrigated with treated 'gray' water and with fresh water. Other studies have reported that the level of fecal indicators in soil irrigated with raw or treated wastewater can significantly differ from that with freshwater application (Malkawi and Mohamad, 2003 and Travis et al., 2010). It also needs to be considered that in the study of Malkawi and Mohamad (2003) there were no fecal coliforms recorded in the soil before the irrigation with fresh water, and thus sources other than water can affect this indicator. With Travis et al. (2010), the levels of fecal coliforms in soil irrigated with untreated or treated gray water was always below 100 CFU g^{-1}.

Considering that *E. coli* was not isolated from any of the soil samples in the present study, it is possible that the die-off, or at least the loss of cultivability, of this important indicator in the present field study occurred faster than previously reported (Lang et al., 2007 and Van Elsas et al., 2011). However, it is more likely that the levels of viable and cultivable *E. coli* under these given environmental conditions were below the sensitivity of the method applied (10^2 CFU g^{-1}) (Samarajeewa et al., 2010). Analogous data were obtained for the plant and fruit here, as no *E. coli* was isolated from either, and the levels of fecal coliforms and THC were relatively low and not influenced by the water used for irrigation ($P > 0.05$). These data are in agreement with other studies (Cirelli et al., 2012), in which only fruit samples (tomato and eggplant) directly in contact with the soil where contaminated by fecal bacteria. Wood et al. (2010) showed that the decline in *E. coli* on the surface of spray-irrigated spinach was considerably rapid (3–5 log reduction in 72 h, and no isolation after 6 days). Another study reported that in the summer months, which are characterized by higher sunlight exposure of the crops, there was a more rapid decay of both the indicator and pathogenic microorganisms (Sidhu et al., 2008). In the present study, there was no particular increase in the microbial contamination for the TW-treated tomato crops. We would thus argue that

the good microbial quality of tomato fruits, (no *E. coli*, very low fecal coliforms, no*Salmonella* spp.), can be seen as the positive consequence of several factors: among the principal ones, the drip irrigation system, the summer month with increased UV radiation exposure of fruit and leaves surfaces. Also, the interval time between irrigation and sampling may have contributed to reduce the effect of treated wastewater on the microbial load of fruits.

3.3.1. Effects on soil microbial communities

Apart from the possible contamination by fecal microorganisms, our aim was also to determine the short-term impact of irrigation with wastewater on the soil microbial communities of this tomato cropping system. The CCA of the ARISA matrix show many interesting aspects related to the dynamics of the bacterial communities in the soil (Fig. 3).

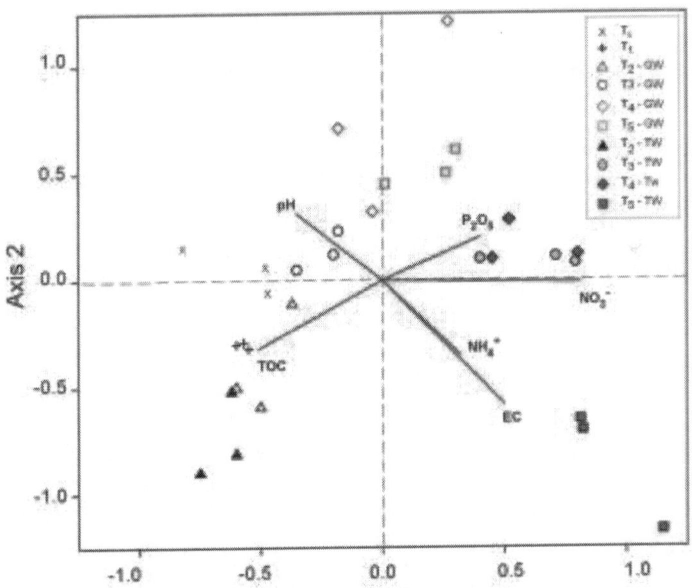

Figure 3. Canonical correspondence analysis base on the ARISA matrix and the chemical soil parameters. For abbreviations, see main text. T_0, sampling before transplanting; T_1, sampling 15 days after transplanting and before the water irrigation treatments; T_2, T_3, T_4 and T_5, sampling at 42, 69, 96 and 124 days after transplanting, respectively.

There was a partial separation of the T_0 samples (soil sampled 1 week before transplanting) from the later time points. The T_1 (15 days after transplanting, but before starting GW and TW irrigation) and T_2 (42 days after transplanting; 27 days of GW or TW irrigation) time points clustered together, with the T_0 samples nearby. Both the T_1 and T_2time points and T_0 were affected by TOC, but not by ammonia or phosphorous. The partial shift from T_0 to the T_1/T_2 clusters might indicate a rhizosphere effect on the soil microbial community, and that this effect is stronger than the irrigation methodology. Hence, this would mean that in the early stages of cultivation, the root exudates of tomato plant are the more important factor in shaping the bacterial communities, with respect to the water treatments used. Indeed, root exudates are known to contain compounds that can exert stimulatory and inhibitory influences on rhizosphere microbial community structure and composition (Dennis et al., 2010 and Hartmann et al., 2009).

Even if soil pH is generally considered a major factor in controlling the soil microbial diversity and composition across a wide range of habitats (Fierer and Jackson, 2006), in the present study, pH did not appear to be an important contributor to the microbial community structures, which is also in agreement with other studies on tomato cropping systems (Buyer et al., 2010).

After almost a month of GW or TW irrigation, at T_2, the first separation between the GW and TW samples is seen, which suggests that these different irrigation waters do affect the microbial communities, and even after this relatively short irrigation period. From T_3(69 days after transplanting; 54 days of GW or TW irrigation), there is a clear and significant separation between GW and TW, and this separation becomes stronger as time progresses. The TW samples show the greater effects, which are evidently correlated to nitrogen (both as ammonia and nitrates), phosphorous and the water conductivity.

The diverging ARISA patterns of the microbial communities after >30 days of irrigation with these different water sources (i.e., from T_2) is in agreement with other studies.Mosse et al. (2012), suggested that it takes >16 days for the effects of the water application on the soil microbial community to become apparent. Despite the strong differences in the fecal indicators in the GW and TW waters and the marked shift of the soil microbial communities, the microbiological quality of the final product (and of the plant) were not significantly different. In a recent study, Telias et al. (2011) investigated the bacterial community diversity and variation on the surface of tomato fruit by culture-independent methods (i.e., pyrosequencing). They found that even if water with very different microbial characteristics is used for long periods of spraying of the tomato surfaces, this did not have any significant impact on the bacterial composition of the tomato fruit surface. Another study aimed to evaluate the potential transfer of enteric bacteria from the soil to the plants by soil splash created by rain-sized droplets (Monaghan and Hutchison, 2012). In this case, both the transfer and survival of artificially inoculated human pathogens occurred, even if the persistence was considerably reduced during the summer months. In the present study, notwithstanding the high input of enteric bacteria into soil, when TW was used, the microbiological quality of final product was not compromised.

4. CONCLUSIONS

In the present study, we wanted to evaluate the use of agro-industrial wastewater in a closed circle system where an agriculture and food manufacturing company produces and processes tomatoes. The aim was to determine whether TW from the company can be used for tomato cropping without compromising the quality and safety of the final product.

Our data have revealed many interesting aspects: (i) the yields of the tomato fruit irrigated with TW were not significantly different from those obtained when the crop was irrigated with GW; (ii) for both the GW and TW irrigation treatments, the most important morpho-qualitative parameters of the processing tomato fruit (i.e., dry matter content, pH, soluble solid content, color parameters) were in agreement with those reported in the literature; and (iii) tomatoes microbial quality was very good for all the thesis stated, even when treated agro-industrial wastewaters were used. That was made possible by combining an accurate control of irrigation treatments with good agricultural practices.

Even if all of the fecal indicators monitored were well over the threshold of the Italian legislation limit for irrigation re-use of TW, the possible die-off of *E. coli* in the soil and the low levels of total coliforms in the soil and plants are of particular interest. To a broader extent, the community composition and the dynamics of the whole bacterial population in the soil was affected by the different qualities of the GW and TW that were used for the tomato crop irrigation.

The present study focused on a comprehensive multidisciplinary approach for the assessment of product quality and safety during a single crop cycle, to evaluate the short-term effects on the use of TW from the food industry. These data are encouraging, even though they are based on a relatively short period of observation. Further studies are being planned to determine the mid-to-long-term effects in the same experimental field.

ACKNOWLEDGMENTS

This study was part of the Project "Technology and process innovations for irrigation re-use of treated municipal and agro-industrial wastewaters in order to achieve sustainable water resources management" (In.Te.R.R.A.—contract no. 01-01480), co-funded by the Italian Ministry of Universities and Research (MIUR), within the Italian "PON/Ricerca e Competitività 2007–2013" Programme.

REFERENCES

1. Aiello, R., Cirelli, G.L., Consoli, S., 2007. Effect of reclaimed wastewater irrigation on soil and tomato fruits: a case study in Sicily (Italy). Agric. Water Manage. 93, 65–72.
2. Al-Lahham, O., El Assi, N.M., Fayyad, M., 2003. Impact of treated wastewater irrigation on quality attributes and contamination of tomato fruit. Agric. Water Manage. 61, 51–62.
3. Allen, R.G., Pereira, L.S., Raes, D., Smith, M., 1998. Crop evapotraspiration guidelines for computing crop water requirements. In: Irrigation and Drainage Paper No. 56. Food and Agriculture Organisation of the United Nations (FAO), Rome.
4. APAT, IRSA-CNR, 2003. Metodi analitici per le acque. Manuali e linee guida. APAT, IRSA-CNR.
5. APHA, AWWA, WEF, 2005. Standard Methods for the Examination of Water and Wastewater, XX ed. American Public Health Association, Washington, DC. Ayers, R.S., Westcot, D., 1985. Water quality for agriculture. In: FAO Irrigation and Drainage Paper No. 29. Food and Agriculture Organisation of the United Nations (FAO), Rome.
6. Benami, M., Gross, A., Herzberg, M., Orlofsky, E., Vonshak, A., Gillor, O., 2013. Assessment of pathogenic bacteria in treated graywater and irrigated soils. Sci. Total Environ. 458–460, 298–302.
7. Blumenthal, U.J., Mara, D.D., Peasey, A., Ruiz-Palacios, G., Scott, R., 2000. Guidelines for the microbiological quality of treated wastewater used in agriculture: recommendations for revising WHO guidelines. Bull. World Health Organ. 78 (9), 1104–1116.

8. Bohner, J., Bangerth, F., 1988. Cell number, cell size and hormone levels in semiisogenicmutants of Lycopersicon pimpinellifoliumdiffering in fruit size. Physiol. Plant. 72, 316–320.

9. Buyer, J.S., Teasdale, J.R., Roberts, D.P., Zasada, I.A., Maul, J.E., 2010. Factors affecting soil microbial community structure in tomato cropping systems. Soil Biol. Biochem. 42, 031–841.

10. Caliandro, A., Lamaddalena, N., Stelluti, M., Steduto, P., 2005. Caratterizzazione agroecologica della Regione Puglia in funzione della potenzialità produttiva. Progetto ACLA. Ideaprint, BARI, Italy, ISBN: 2-85352-339-X.

11. Cardinale, M., Brusetti, L., Quatrini, P., Borin, S., Puglia, A.M., Rizzi, A., Zanardini, E., Sorlini, C., Corselli, C., Daffonchio, D., 2004. Comparison of different primer sets for use in automated ribosomal intergenic spacer analysis of complex bacterial communities. Appl. Environ. Microbiol. 70, 6147–6156.

12. Chen, W., Wu, L., Frankenberger Jr., W.T., Chang, A.C., 2008. Soil enzyme activities of long-term reclaimed wastewater-irrigated soils. J. Environ. Qual. 37, 36–42 (September–October Supplement).

13. Cirelli, G.L., Consoli, S., Licciardello, F., Aiello, R., Giuffrida, F., Leonardi, C., 2012. Treated municipal wastewater reuse in vegetable production.Agric.Water Manage. 104, 163–170.

14. Cooley, W.W., Lohnes, P.R., 1971. Multivariate Data Analysis. Wiley, New York, NY. Decree of Ministry for Environment, 2003. Decree of Ministry for Environment, No. 185, 12/06/2003, Gazzetta Ufficiale, No. 169, Rome July 23, 2003. Italian Technical Guidelines for Wastewater Reuse.

15. Dennis, P.G., Miller, A.J., Hirsch, P.R., 2010. Are root exudates more important than other sources of rhizodeposits in structuring rhizosphere bacterial communities? FEMS Microbiol. Ecol. 72, 313–327.

16. FAO, 2010. The wealth of waste: The economics of wastewater use in agriculture. In: FAO Water Report No. 35. Water Development and Management Unit, Food and Agriculture Organization of the United Nations, ISBN 978-92-5-106578-5. FAO, 2011. The State of the World's Land and Water Resources for Food and Agriculture. The Food and Agriculture Organization of the United Nations And Earthscan, ISBN 978-1-84971-326-9 (hdk).

17. Favati, F., Lovelli, S., Galgano, F., Miccolis, V., Di Tommaso, T., Candido, V., 2009. Processing tomato quality as affected by irrigation scheduling. Sci. Hortic. 122 (November (4)), 562–571.

18. Fierer, N., Jackson, R.B., 2006. The diversity and biogeography of soil bacterial communities. Proc. Natl. Acad. Sci. U.S.A. 103, 626–631.

19. Francis, F.J., Clydesdale, F.M., 1975. Food Colorimetry: Theory and Applications. AVI Publ. Co., Westport, CT, pp. 477.

20. Garcia, E., Barret, D.M., 2006. Evaluation of processing tomato from two consecutive growing seasons: quality attributes, peelability and yield. J. Food Process. Preserv. 30, 20–36.

21. Gittins, R., 1985. Canonical Analysis. A Review with Application in Ecology. Stringer, Berlin. Guil-Guerrero, J.L., Rebolloso-Fuentes, M.M., 2008. J. Food Compos. Anal. 22, 123.

22. Hammer, Ø., Harper, D.A.T., Ryan, P.D., 2001. PAST: paleontological statistics software package for education and data analysis. Palaeontol. Electron. 4, 1–9.

23. Hartmann, A., Schmid, M., van Tuinen, D., Berg, G., 2009. Plant-driven selection of microbes. Plant Soil 321, 235–257.

24. Lang, N.L., Bellett-Travers, M.D., Smith, S.R., 2007. Field investigations on the survival of Escherichia coli and presence of other enteric micro-organisms in biosolidsamended agricultural soil. J. Appl. Microbiol. 103, 1868–1882.

25. Lonigro, A., Rubino, P., Brandonisio, O., Spinelli, R., Pollice, A., Laera, G., 2007. Vegetable crops irrigation with tertiaryfilteredmunicipal wastewater. Plant Biosyst. 141 (2), 275–281.

26. Lopez, A., Pollice, A., Laera, G., Lonigro, A., Rubino, P., 2010. Membrane filtration of municipal wastewater effluents for implementing agricultural reuse in Southern Italy. Water Sci. Technol. 625, 1121–1128.

27. Lopez, A., Pollice, A., Laera, G., Lonigro, A., Rubino, P., Passino, R., 2007. Reuse of membrane filtered municipal wastewater for irrigating vegetable crops. In: Proc. of the International Conference on "Water Saving in Mediterranean Agriculture & Future Research Needs, Valenzano (Bari) 14/17 febbraio 2007. Vol. 2, pp. 181–189.

28. Jimenez-Cuesta, M., Cuquerella, J., Martinez–Javaga, J.M., 1981. Determination of color index for citrus fruits degreening. Proc. Int. Soc. Citriculture 2, 750–753.

29. Keeney, D.R., Nelson, D.W., 1982. Nitrogen inorganic forms. In: Methods of Soil Analysis. Part 2, second ed. Agron Monogr 9 ASA and SSSA, Madison, WI, USA, pp. 643–698, 1982.

30. Madrid, R., Barba, E.M., Sanchez, A., Garcia, A.L., 2009. Effects of organic fertilizers and irrigation level on physical and chemical quality of industrial tomato fruit (cv. Nautilis). J. Sci. Food Agric. 89, 2608–2615.

31. Mahajan, G., Singh, K.G., 2006. Response of greenhouse tomato to irrigation and fertigation. Agric. Water Manage. 84, 202–206. Malkawi, H.I., Mohamad, M.J., 2003. Survival and accumulation of microorganisms in soils irrigated with secondary treated wastewater. J. Basic Microbiol. 43, 47–55.

32. Meli, S., Porto, M., Belligno, A., Bufo, S.A., Mazzatura, A., Scopa, A., 2002. Influence of irrigation with lagooned urban wastewater on chimica and microbiological soil parameters in a citrus orchard under Mediterranean condition. Sci. Total Environ. 285, 69–77.

33. Monaghan, J.M., Hutchison, M.L., 2012. Distribution and decline of human pathogenic bacteria in soil after application in irrigation water and the potential for soil-splash-mediated dispersal onto fresh produce. J. Appl. Microbiol. 112, 1007–1019.

34. Mosse, K.P.M., Patti, A.F., Smernik, R.J., Christen, E.W., Cavagnaro, T.R., 2012. Physicochemical andmicrobiological effects oflong- and short-termwinery wastewater application to soils. J. Hazard. Mater. 201–202, 219–228.

35. Nadkarni, M.A., Martin, F.E., Jacques, N.A., Hunter, N., 2002. Determination of bacterial load by real-time PCR using a broad-range (universal) probe and primers set. Microbiology 148, 257–266.

36. AOAC, 1990. Official Method of and Analysis. Official Method of Analysis. Association of Official Analytical Chemists, Washington, DC (No. 934.06). AOAC, 1995. Official Methods of Analysis, 16th ed. Association of Official Analytical Chemists, Washington, DC, pp. 16.

37. Olsen, S.R., Cole, C.V., Watanabe, F.S., Dean, L.A., 1954. Estimation of Available Phosphorus in Soil by Extraction with Sodium Bicarbonate, 939. USDA Circular, Washington, DC, pp. 1–19.

38. Palese, A.M., Pasquale, V., Clano, G., Figliuolo, G., Masi, S., Xiloyannis, C., 2009. Irrigation of olive groves in Southern Italy with treated municipal wastewater: effects on microbiological quality of soil and fruits. Agric. Ecosyst. Environ. 129, 43–51.

39. Patanè, C., Tringali, S., Sortino, O., 2011. Effect of deficit irrigation on biomass, yield, water productivity an fruit quality of processing tomato under semi-arid Mediterranen climate conditions. Sci. Hortic. 129, 540–596.

40. Pedrero, F., Kalavrouziotis, I., Alarcón, J.J., Koukoulakis, P., Asano, T., 2010. Use of treated municipal wastewater in irrigated agriculture—review of some practices in Spain and Greece. Agric. Water Manage. 97, 1233–1241.

41. Podani, J., 2007. Analisi ed esplorazione multivariata dei dati in ecologia e biologia. Liguori, Napoli, Italy.

42. Pollice,A., Lopez,A., Laera, G., Rubino, P., Lonigro,A., 2004. Tertiary filtered municipal wastewater as alternative water source in agriculture: a field investigation in Southern Italy. Sci. Total Environ. 324, 201–210.

43. Patterson, S.R., Ashbolt, N.J., Sharma, A., 2011. Microbial Risks from wastewater irrigation og salad crops: a screening-level risk assessment. Water Environ. Res. 73 (6), 667–672.

44. Rees, G.N., Baldwin, D.S., Watson, G.O., Perryman, S., Nielsen, D.L., 2004. Ordination and significance testing of microbial community composition derived from terminal restriction fragment length polymorphisms: application of multivariate statistics. Antonie van Leeuwenhoek 86, 339–347.

45. Reg. CE n. 1881/2006 del 19/12/2006 in materia di "Tenori massimi di alcuni contaminanti nei prodotti alimentari".

46. Reg. UE n. 1258/2011 del 2 December 2011. GU L 320/15.

47. Renna, M., Gonnella, M., Giannino, D., Santamaria, P., 2013. Quality evaluation of cook-chilled chicory stems (Cichorium intybus L., Catalogna group) by conventional and sous vide cooking methods. J. Sci. Food Agric.,

48. Richardson, C., Hobson, G.E., 2006. Compositional changes in normal and mutant tomato fruit during ripening and storage. J. Sci. Food Agric. 40, 245–252.

49. Richards, L.A., 1954. Diagnosis and Improvement of Saline and Alkali Soils. USDA Agriculture Handbook 60, Washington DC. Rubino, P., Lonigro, A., 2008. Municipal treated wastewater irrigation: microbiological risk evaluation. Ital. J. Agron. 1 (1), 119–124.

50. Sadocchi, S., 1981. Manuale di analisi statistica multivariata per le scienze sociali. Franco Angeli Editore, Milano, Italy, pp. 274.

51. Salgot, M., Huertas, E., Weber, S., Dott, W., Hollender, J., 2006. Wastewater reuse and risk: definition of key objectives. Desalination 187, 29–40.

52. Samarajeewa, A.D., Glasauer, S.M., et al., 2010. Evaluation of Petrifilm EC method for enumeration of E. coli from soil. Lett. Appl. Microbiol. 50 (5), 457–461.

53. Sawhney,V.K., Greyson, R.I., 1972. Fruit size increase in tomato following application of gibberellic acid. J. Am. Soc. Hortic. Sci. 97, 589–590.

54. Sgherri, C., Navari-Izzo, F., Pardossi, A., Soressi, G.P., Izzo, R., 2007. The influence of diluted seawater and ripening stage on the content of antioxidants in fruits of different tomato genotypes. J. Agric. Food Chem. 55, 2452–2458.

55. Sgherri, C., Kadlecova, Z., Pardossi, A., Navari-Izzo, F., Izzo, R., 2008. Irrigation with diluted seawater improves the nutritional value of cherry tomatoes. J. Agric. Food Chem. 56, 3391–3397.

56. Sidhu, J.P., Hanna, J., Toze, S.G., 2008. Survival of enteric microorganisms on grass surface irrigated with treated effluent. J. Water Health. 6 (2), 255–262.

57. Siebe, C., Cifuentes, E., 1995. Environmental impact of wastewater irrigation in central Mexico: an overview. Int. J. Environ. Health Res. 5, 161–173.

58. Soraya, G., Nadia, B., Cherubino, L., Christian, G., 2001. Tomato fruit quality in relation to water and carbon fluxes. Agronomie 21, 385–392.

59. Suárez, M.H., Rodríguez, E.M.R., Romero, C.D., 2007. Food Chem. 104, 489.

60. Telias,A.,White, J.R., Pahl, D.M., Ottesen,A.R.,Walsh, C.S., 2011. Bacterial community diversity and variation in spray water sources and the tomato fruit surface. BMC Microbiol. 11, 81.

61. Toze, S., 2006. Reuse of effluent water-benefits and risks. Agric. Water Manage. 80, 147–159.

62. Travis, M.J., Wiel-Shafran, A., Weisbrod, N., Adar, E., Gross, A., 2010. Greywater reuse for irrigation: effect on soil properties. Sci. Total Environ. 408, 2501–2508.

63. Turhan, A., Seniz, V., 2009. Estimation of certain chemical constituents of fruits of selected tomato genotypes grown in Turkey. Afr. J. Agric. Res. 4, 1086–1092. UNI EN ISO 19250:2013. Water quality-Detection of Salmonella spp. (ISO 19250:2010).

64. Van Elsas, J.D., Semenov, A.V., Costa, R., Trevors, J.T., 2011. Survival of Escherichia coli in the environment: fundamental and public health aspects. ISME J. 5, 173–183.

65. Vivaldi, G.A., Camposeo, S., Rubino, P., Lonigro, A., 2013. Microbial impact of different types of municipal wastewaters used to irrigate nectarines in Southern Italy. Agriculture. Ecosyst. Environ. 181, 50–57.

66. Walkley, A., Black, I.A., 1934. An examination of the Degtjareff method for determining organic carbon in soils: effect of variations in digestion conditions and of inorganic soil constituents. Soil Sci. 63, 251–263. WCED, 1987.

67. Our Common Future.World Commission on Environmental and Development. Oxford University Press, Oxford, UK, pp. 1987. Wood, J.D., Bezanson, G.S., Gordon, R.J., Jamieson, R., 2010. Population dynamics of Escherichia coli inoculated by irrigation into the phyllosphere of spinach grown under commercial production conditions. Int. J. Food Microbiol. 143, 198–204.

68. World Health Organization (WHO), 2006. Guidelines for the Safe Use of Wastewater, Excreta and Greywater, Volume 4. Excreta and Greywater Use in Agriculture. World Health Organization, Geneva.

11
CHAPTER

Spring Wheat Response to Disease Control and Subsurface Drainage Management in the Red River of the North Valley, USA

Grant H. Mehring[1*], Hans J. Kandel[1], Joel K. Ransom[1], Amanda Schoch[1], Dean D. Steele[2]

[1]*Department of Plant Science, North Dakota State University, Fargo, ND, USA*

[2]*Department of Agricultural and Biosystems Engineering, North Dakota State University, Fargo, ND, USA*

ABSTRACT

Increased variability in rainfall events and high production input costs are driving agricultural producers to consider subsurface water management in the flat Red River of the North Valley in Eastern North Dakota and Northwestern Minnesota, USA. Subsurface tile incorporated with water table control structures was utilized from 2009 to 2011 to investigate the response of hard red spring wheat (HRSW) (Triticum aestivum L. emend. Thell.) for yield, disease, and other agronomic characteristics to soil water management. A factorial arrangement of four cultivars, two seed treatments, and two foliar fungicide treatments in a split-plot design with closed and open tile as whole-plots was used. Mean wheat yields averaged across years were not significantly different with closed or open tile treatments. There existed an optimum management practice where plant useable water was not freely drained and analyzing the data with the optimum water management for each year found the optimum water table managed treatment yielded higher with 3812 kg ha^{-1} compared with limited water table management with 3679 kg ha^{-1}. In 2011, the cultivars Faller and Howard were taller, and Traverse had lower root disease severity. In 2010 and 2011, Howard and Traverse had more leaf disease with open tile compared with closed tile. Across years, there was no difference in root disease, stand, number of spikes, crop height, or yield response to application of seed treatments with open or closed tile. In 2010, there was a 3.7% yield advantage with application of seed treatment on open tile. Across years, there was no yield response to application of foliar fungicides; however, wheat yield with foliar fungicide was 5% higher than without application in 2010. Producers should be using water table control and disease management to maximize HRSW yield. Further research should investigate water table management throughout the season based on weather conditions.

Keywords: Spring Wheat, Disease Control, Subsurface Drainage, Water Table

1. INTRODUCTION

Many different crops are grown in eastern North Dakota, in the USA, in an area known as the Red River Valley (RRV). The region is characterized as having a cold, sub-humid continental climate [1] . The RRV has level topography, and poorly-drained, clay-rich soils [2] . Fertile soil and favorable weather conditions have allowed North Dakota to occasionally rank first in the national production of wheat across market classes [3] . However, since the early 1990s the RRV has been in a wet climatic cycle, which has caused a decrease in crop yield potential due to increased disease pressure [4] .

Surface drainage utilizing field ditches is the main method to alleviate excessive moisture conditions in the RRV [2] . In recent years the interest in subsurface tile drainage has increased, due to the wet climatic cycle and to help reduce the financial risk associated with high crop production input costs to farmers. Subsurface drainage tends to result in more stable production at a higher yield level [2] .

For years subsurface drainage has been a common practice in many parts of the Upper Midwest, in the USA, but adoption of subsurface drainage into the RRV is relatively new. The slow adoption is largely due to the flat topography and limited natural outlets for subsurface drained systems [2] . Using subsurface drainage helps lower the water table and provides better soil aeration [5] . A lower water table causes soils to warm-up faster and potentially allows for earlier planting dates in cold and humid climates such as in the RRV. Soil temperature impacts seed germination, emergence, seedling growth, and soil microbial activity [6] . Subsurface drainage has also helped reduce soil compaction and decrease water-logging stress to plants [5] . Research has shown that subsurface drainage systems can increase grain yields, and improve soil trafficability and workability [7] [8] .

In the USA there is an increased interest in managing the water table in subsurface drainage systems with control structures. Research on maize (Zea mays (L.)) and soybean (Glycine max (L.) Merrill) in Missouri has shown positive yield responses to controlled drainage and sub-irrigation [9] - [12] . However, research is still needed to address the agronomic impacts of subsurface drainage management on HRSW in the climatic and soil conditions of the RRV.

In susceptible wheat cultivars, fungal leaf spot diseases can cause reduced test weights and yield losses up to 50 percent [13] . Fusarium head blight (FHB), [Fusarium graminearum (Schwabe)], or scab, is another fungal disease that can cause significant yield loss in wheat [14] in addition to quality reductions [15] .

Water-logged soils may cause root and crown rot in small grains [16] [17] . Phythium root rot (Pythium spp.) and take-all [Gaeumannomyces graminis (Sacc.)] are common in North Dakota. To help reduce the risk of crop injury and yield loss due to disease pressure, there are a number of effective management tools including crop rotation, residue management [13] , seed treatments [18] [19] , efficient timing of systemic fungicides [15] , use of resistant cultivars [15] , and possibly subsurface water management. Multiple tools should be utilized in a plant disease management strategy.

Studies conducted on hard red winter wheat observed that the application of fungicides at flowering (Zadoks 58) [20] improved yield and grain quality [15] . The yield increase was associated with the management of foliar leaf disease and FHB. The use of fungicides was profitable when applied to cultivars under high disease pressure, but not profitable when disease levels were low. Results showed that the combination of resistant varieties and fungicide application resulted in the highest yield and the greatest financial return [15] .

With the wetter climatic cycle in the Northern USA and an increase in subsurface drainage management, producers are seeking additional management techniques to successfully grow HRSW. Previous researchers have not investigated the effect of subsurface drainage management, seed treatment, foliar application of fungicides and their interactions on yield, disease, and other agronomic characteristics in HRSW. The objectives of this research were to evaluate several HRSW cultivars' response to water table management with seed treatments, foliar applications of fungicides, and their interactions on selected plant characteristics, yield, and disease development when grown near Fargo.

2. MATERIALS AND METHODS

2.1. Regional and Site Description

The experiment was conducted in 2009, 2010, and 2011, at 46.932°N and 96.858°W, near Fargo, ND. This site is situated in the RRV, which was once part of the glacial Lake Agassiz. The predominant soil at this site is Fargo-Ryan silty clay. The Fargo series (fine, smectic, frigid, Typic Epiaquerts) consists of deep, poorly drained, slowly permeable, lacustrine soils. This soil generally has a slope of 0% to1%. The Ryan series (fine, smectic, frigid, Typic Natraquerts) is very similar to the Fargo series [21] . Soil analysis before the study began in 2009 for the 0 - 15 cm depth was 42 ppm NO_3-N, 13 ppm P, 426 ppm K, 7.7 pH and 4.7% OM while the 15 - 70 cm depth had 563 ppm NO_3-N, 7 ppm P, 339 ppm K, 7.9 pH and 3.4% OM. The region has an annual average precipitation of 533 mm with 64% of that occurring as rainfall from May to September [6] . Each year the previous crop was soybean.

The experimental design was a randomized complete block with four replicates in a split-plot arrangement where the main plots were open and closed subsurface tile drainage management and the sub-plots a 4 × 2 × 2 factorial combination of cultivar, seed treatment, and in-season fungicide application.

The drainage treatment consisted of eight 0.33-ha land units with independent subsurface water control structures (Agri Drain Corp, Adair, IA), which can set the drainage outlet height, and in effect the water table as described by Kandel et al. [8] . There were seven 102 mm diameter horizontal subsurface tile lines for each land unit with a spacing of 7.6 m apart and 100 cm tile depth and a main outlet to a drainage ditch located to the west of the units. Of these eight control structures four were kept closed (tile closed) to simulate non-subsurface drained conditions, while the other four were left open (tile open) to represent subsurface drained soil. The whole plot of open and closed subsurface tile drainage was analyzed statistically in two ways. Management level 1 was open and closed tile while management level 2 was best water table management practice. The outlet levels were set at 100, 90, and 85 cm below the surface in 2009, 2010, and 2011, respectively. Corresponding with a drainage coefficient of 8.4, 7.0, and 6.4 mm per 24 h for the 100, 90, and 85 cm below the surface settings, respectively. The outlet level in 2010 and 2011 was set closer to the surface based on the experimental results from 2009.

The wheat experimental plots (width 1.5 m and length 7.6 m) consisted of seven rows at 18-cm row spacing and were seeded perpendicular to the tile lines and the center of the plot was on top of the tile line. Alleys 0.8-m wide were later cut on the long ends between plots. The total harvested plot area

Table 1. Daily average temperature and precipitation for each month during the growing season in 2009 to 2011, Fargo, ND.

Month	Max Air Temp				Min Air Temp				Precipitation			
	2009	2010	2011	Avg.†	2009	2010	2011	Avg.	2009	2010	2011	Avg.
	-------°C-------				-------°C-------				-------mm-------			
April	10	18	11	13	1	4	2	0	16	37	46	35
May	19	20	18	21	5	9	8	7	44	68	110	66
June	24	25	24	25	12	14	14	13	82	86	101	89
July	26	28	29	28	13	17	18	15	16	105	104	73
August	25	28	28	27	13	16	16	14	47	68	73	64
Total									205	364	434	327

† Avg. is 30-year average (1971-2000) [11].

was 1.5 m × 5.8 m. The seeding rate was 3 million pure live seeds ha^{-1}. Two rain gauges were stationed at the research site to measure daily rainfall. Daily average temperature and precipitation data for each month during the growing season were collected by the Fargo automatic weather station 5 km away that is part of the North Dakota Agricultural Weather Network (Table 1) [1] .

2.2. Plant Materials

Four HRSW cultivars and buffer strips ("Glenn"), on each side of the research area, were planted as early in the season as possible, using a Great Plains 3P605NT no-till plot drill (Great Plains Mfg Inc., Salina, KS). Cultivars were selected based on their variable response to leaf disease, FHB, and other characteristics (Table 2) [22] [23] .

Fertilizer P and K was applied in the fall the year before seeding based on soil tests. In addition fertilizer N was applied as urea (46-0-0) at a rate of 112 kg · N · ha^{-1} in the spring using a Gandy 10T Series fertilizer spreader (Gandy Company, Owatonna, MN), for a yield goal of 4500 kg ha^{-1}[24] . In 2010 an additional 56 kg · N · ha^{-1} as urea was surface applied with a Gandy 1.2 m Nursery Special (Gandy Company, Owatonna, MN). This second application was done 07 June at the boot stage (Zadoks 45) because signs of leaf yellowing were present. Dates for field operations are provided in Table 3.

Before planting, seed receiving the seed treatment were treated with Tebuconazole: Alpha-[2-(4-chlorophe- nyl)ethyl]-alpha-(1,1-dimethyl-ethyl)-1H-1,2,4-triazole-1-ethanol and Metalaxyl: N-(2,6-dimethylphenyl)-N- (methoxyacetyl) alanine methyl ester; formulated as Raxil MD, (Bayer CropScience, Research Triangle Park, NC) applied at rate of 3.78 g/ai 100 kg^{-1}. Seed was treated using a Hege II batch seed treatment machine (Waldenburg, Germany). Foliar fungicide in-season treatment consisted of no fungicide or fungicide at anthesis (Zadoks 60). The fungicide used was Prothioconazole: 2-[2-(1-Chlorocyclopropyl)-3-(2-chlorophenyl)-2-hydroxy- propyl]-1,2-dihydro-3H-1,2,4-triazole-3-thione and Tebuconazole: alpha-[2-(4-chlorophenyl)ethyl]-alpha-(1,1- dimethylethyl)-1H-1,2,4-triazole-1-ethanol; formulated as Prosaro 421 SC, (Bayer CropScience, Research Triangle Park, NC) at rate of 252.2 g · ai · ha^{-1} in 93.5 L ha^{-1} waterat 200 kPa using XR TEEJET 8001 VS nozzles. North Dakota Extension Service recommended fungicide application techniques were used [25] . Weeds were controlled with an application ofFenoxaprop-p-Ethyl, Pyrasulfotole, Bromoxynil Octanoate, and Bromoxynil Heptanoate; formulated as Wolverine (Bayer CropScience, Research Triangle Park, NC)

Table 2. Characteristic of hard red spring wheat cultivars used in 2009 to 2011, Fargo, ND.

	Height†	Days to Heading	Stem Rust‡	Leaf Rust‡	Foliar Disease‡,§	Fusarium Head Blight‡	Straw Strength¶
	cm	DAP#	1 - 9	1 - 9	1 - 9	1 - 9	1 - 9
Faller††	81	54	1	1	3	4	5
Howard††	84	52	1	1	4	6	7
Sabin‡‡	76	53	1	3	6	4	6
Traverse§§	85	51	2	5	5	5	6

†Sources: [1] [18]. ‡1 - 9 scale where 1 = most resistant, 9 = most susceptible. §Includes tan spot, septoria, bacterial leaf blight, and powdery mildew. ¶1 - 9 scale in which 1 is the strongest straw and 9 the weakest. #DAP is days after planting. ††Variety developed and released by North Dakota Agricultural Experiment Station, including buffer strip variety Glenn. ‡‡Variety developed and released by University of Minnesota Agricultural Experiment Station. §§Variety developed and released by South Dakota Agricultural Experiment Station.

Table 3. Dates of field operations in 2009 to 2011, Fargo, ND.

Operation	2009	2010	2011
Fertilizing	08 May	28 April	19 June
Seeding	11 May	12 April	04 May
Stand count	27 May	04 May	03 June
Herbicide application	09 June	18 May	09 June
Fungicide application	09 July	18 June	04 July
Harvest	24 August	05 August	10 August

at a rate of 2 L ha^{-1} equal to 272 g · ai · ha^{-1}, at the four-leaf stage (Zadoks 14).

2.3. Plant Observations and Disease Assessment

Plant stand was determined at the two-leaf stage (Zadoks 12) by counting a random 30-cm length of row from rows two and six and averaging to determine plants ha^{-1}. Spike count, after anthesis (Zadoks 69) (2010-11 only) was determined by counting spikes in 91 cm of rows two and six and averaging to determine spikes ha^{-1}. Crop height measurement was taken twice per plot and averaged, (2010-11 only) before harvest from ground level to the top of the spike, excluding the awns.

Crown root rot ratings were based on the extent of lesion development on the sub-crown internodes and mesocotyl [26] . Five plants were randomly selected in each plot at the soft dough stage (Zadoks 85) and cleaned by wiping soil away on the mesocotyl. Disease scores were then assigned as a percent of mesocotyl discolored [27] .

Leaf disease notes were taken when plants were at the soft dough stage (Zakoks 85). The combined level of severity of leaf rust and leaf spot diseases (tan spot and septoria) were visually estimated based on percent disease covering the flag leaf. Five random flag leaf samples were evaluated and data averaged.

Fusarium head blight was estimated from ten randomly selected spikes at ripening (Zakoks 90). The intensity was assessed as the number of spikes infected with FHB and the severity as the percentage of the spike damaged by FHB [28] .

Plots were harvested with a small plot combine. The harvested grain was then dried and cleaned with a Clipper office tester and cleaner (Seedburo Equipment Co., Chicago, IL) and weighed. Harvest moisture and test weight were recorded with a Dickey-John GAC 2100 (Dickey-John Corp., Minneapolis, MN). Yield was adjusted to reflect 13% grain moisture content.

2.4. Water Table Depth

Water table depths were measured for a whole plot effect of closed and open tile. Within each of the eight units, four observation wells midway between adjacent subsurface drain lines near the center of each unit were used to measure the water table depth below the soil surface. The wells consisted of vertically installed pipes 122 cm and 213 cm deep. The depth to the water table was measured using a Solinst water level meter (Model 101, Solinst, Georgetown, ON, Canada). Water table depths were recorded from the soil surface to the top of the water table. Recorded water depths were averaged across each drainage treatment.

2.5. Statistical Analysis

For each year, the statistical analysis was performed utilizing standard procedures for a randomized complete block design with a split-plot arrangement for closed versus open tile. If the error mean square between the years was homogenous, then a combined analysis across years was conducted for that trait. In 2009, the closed water structure was holding back water in the middle of July which would have been the best water management practice for that year due to dry growing conditions. For 2010 and 2011 open control structures were the best water management practice due to wetter conditions. These observations led us to consider the two water table management levels for statistical analysis of the yield data combined across years. In management level 1, a combined analysis of closed tile versus open tile treatments was conducted with no consideration of the setting of the water table position in the control structure. In management level 2, we analyzed the best water table management practice compared with the other treatment, where the best management treatment was closed tile in 2009 and open tile in 2010 and 2011. The best water management treatment for level 2 was responsive to in-season weather and water table position.

For the statistical analysis, tile or water management, cultivar, seed treatment, and fungicide were considered fixed effects. Years and replicates were considered random effects and all interactions of fixed effects with year or replicate were considered random. Data analysis was conducted using the mixed model procedure (PROC MIXED) of SAS version 9.3 (SAS Institute, SAS Circle, Cary, NC). Treatment means were separated using the PDIFF function under the LSMEANS statement at an alpha of 0.10 level of significance for each evaluated characteristic.

3. RESULTS AND DISCUSSION

3.1. Rainfall, Temperature and Water Table

3.1.1. 2009

The summer of 2009 was relatively dry (Table 1) with only 63% of the long-term precipitation, and cooler than normal. There were no rainfall events intense or large enough to produce surface runoff events at the trial location. From the 11 May seeding through 01 July the water table was below the tile line. After repeated rainfall events at the end of June the water table rose above the tile line in both treatments. With the open tile there was limited drainage from 01 July through 08 July (early to late boot stage, Zadok 45 - 49). With the control structure closed the water was retained and the water table was 12 cm higher than the water table in the open tile treatment for a portion of July (Figure 1(a)). For the rest of the season the water table was below the tile line.

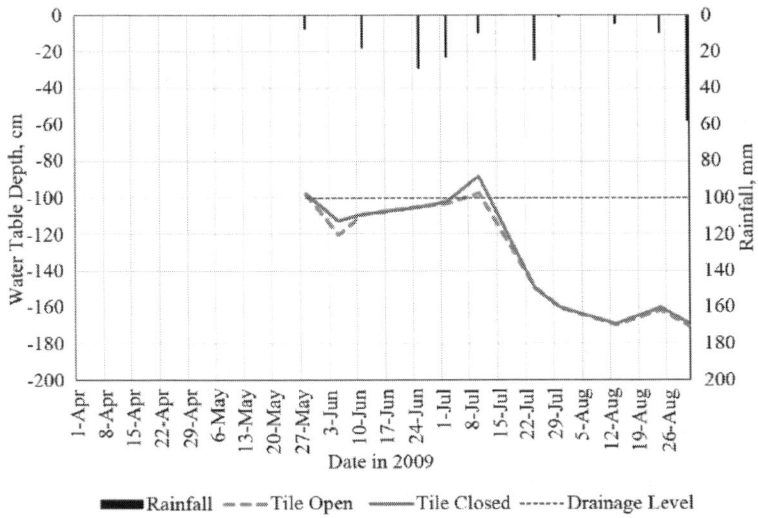

(a)

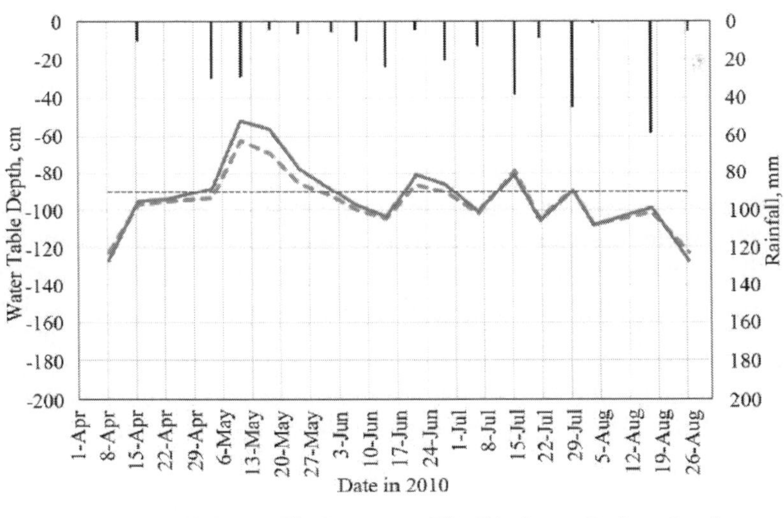

(b)

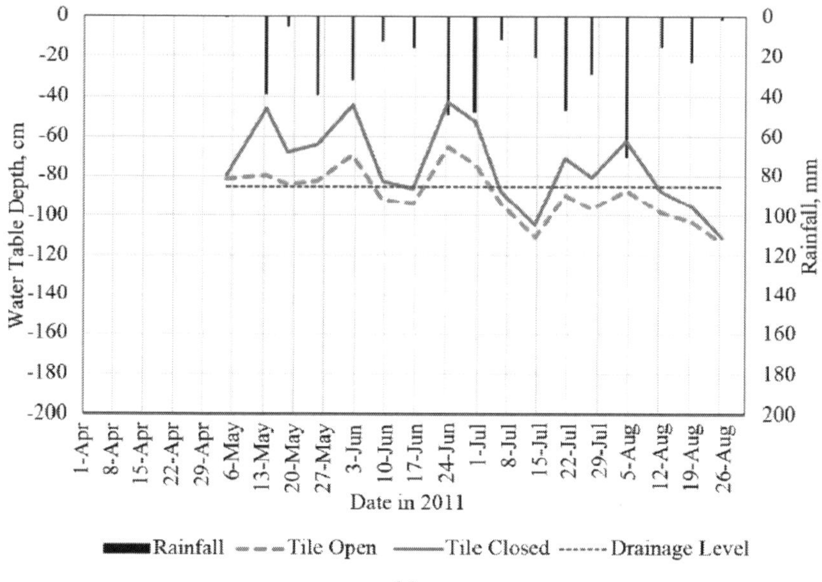

Figure 1. (a) Water table depth and rainfall in 2009, Fargo, ND. (b) Water table depth and rainfall in 2010, Fargo, ND. (c) Water table depth and rainfall in 2011, Fargo, ND.

3.1.2. 2010

The temperature in April was higher than normal and the HRSW was seeded by the middle of April. The average maximum temperature was close to normal but the night (minimum temperatures) was above average during the growing season. Rainfall was close to normal for all months except July when there was 40% more precipitation than normal (Table 1). The water table was below the tile line until about 06 May. The open tile plots were releasing water through 01 June. The water table in the open tile situation was consistently deeper than the closed tile treatment. In the closed tile treatment the water table reached a level of 52 cm below the surface on 10 May (Figure 1(b)) and the wheat was in the three leaf stage (Zadok 13).

3.1.3. 2011

The early summer of 2011 was relatively cool with below average maximum temperatures for April and May. The weather turned hot during flowering and grain fill. Every month of the growing season had higher

Table 4. Probability level of significance for yield with tile open in control boxes versus closed, for individual years, combined 2009 to 2011 and for the best management strategy for water table management combined in 2009 to 2011, Fargo, ND.

					Management	Management
					Level 1[†]	Level 2[‡]
		2009	2010	2011	Closed vs Open 2009-2011	Best Managed 2009-2011
Sources of Variation	df	-----------------	Probability > F	-----------------		
Closed vs Open Tile (Main)	1	0.19	0.74	0.18	0.97	**0.08**[§]
Cultivar	3	**0.01**	**<0.0001**	**0.02**	0.49	0.50
Main × Cultivar	3	0.59	0.52	**0.09**	0.29	0.99
Seed Treatment (ST)	1	0.15	0.74	0.14	0.12	0.12
Main × ST	1	0.22	**0.09**	**0.06**	0.66	0.72
Cultivar × ST	3	0.52	0.70	0.29	0.34	0.34
Main × Cultivar × ST	3	**0.02**	0.57	0.73	0.49	0.46
Foliar Fungicide (FF)	1	0.45	**<0.0001**	0.49	0.41	0.41
Main × FF	1	0.79	0.23	0.53	0.59	0.87

Cultivar × FF	3	0.90	0.40	0.43	0.41	0.41
Main × Cultivar × FF	3	0.62	0.65	0.92	0.92	0.29
ST × FF	1	0.53	**0.07**	**0.07**	**0.05**	**0.05**
Main × ST × FF	1	0.76	0.12	0.24	0.10	0.32
Cultivar × ST × FF	3	0.55	0.43	0.79	0.13	0.13
Main × Cultivar × ST × FF	3	0.81	0.29	0.85	**0.04**	0.61

[†]Level 1 is closed or open tile. [‡]Level 2 is a best management approach, *i.e.*, tile closed in 2009 in response to dry conditions and tile open in 2010 and 2011 in response to wetter conditions. [§]Bold numbers indicate significance at $P \leq 0.10$.

Table 5. Hard red spring wheat yield with closed and open tile as whole plot factors and with limited management and best management as whole plot factors in 2009 to 2011, Fargo, ND.

Management Level 1[†]	2009	2010	2011	2009-2011	Management Level 2[‡]	2009-2011
	------------	kg ha^{-1}	------------			--kg ha^{-1}--
Closed tile	4696 a[§]	4155 a	2378 a	3743 a	Limited management	3679 b
Open tile	4504 a	4211 a	2528 a	3748 a	Best Management	3812 a

[†]Level 1 is closed or open tile. [‡]Level 2 is a best management approach, *i.e.*, tile closed in 2009 in response to dry conditions and tile open in 2010 and 2011 in response to wetter conditions. [§]Numbers within the same column with the same letter are not significantly different at $P \leq 0.10$.

precipitation levels than the normal, resulting in about 30% more rainfall during the growing season (Table 1). The water table with the control structure open was consistently deeper than the water table with closed tile. In the closed situation the water table reached a level of about 44 cm below the soil surface (Figure 1(c)) on 13 May (plants just emerged, Zadok 10), 13 June (four leaf stage, Zadok 14) and June 24 (early boot, Zadok 44 - 25).

3.1.4. Yield

Wheat yield with open tile was not significantly different (at $P \leq 0.10$) from the yield with the control structure closed (Table 4 and Table 5). This corresponds with management level 1 where no adjustment of the control structures was made in response to precipitation or water table. These results are similar to equal HRSW yields reported by Wiersma et al. [2] for tiled and non-tiled treatments and equal winter wheat yields reported by Poole et al. [29] . Evaluating the individual years we noticed that in 2009 yield with closed tile was numerically higher than the open tile, but the opposite in 2010 and 2011. Poole et al. [29] reported non-significant increase in winter wheat yields in 2 out of 4 environments with controlled drainage versus free drainage. They concluded that yield gains were higher when water was beneficially held back during dry periods and losses were found when water was held back past the optimum water level. The data from Wiersma et al. [2] indicated a non-significant numerical HRSW yield increase of about 4% with drained plots with drainage coefficients of 6.3 and 12.7 mm/d^{-1} compared to the control (no subsurface drainage) in seven environments over four years. However, a 19 mm/d^{-1}coefficient resulted in identical yield with the control. This seems to indicate that plant useable water removal may negate potential positive effects of drainage. With excess moisture at the wrong growth stage crop yields may be decreased but there also may be periods in the growing season that a crop benefits from additional stored water via controlled drainage [29] .

During 2009 there was some water kept above the tile line with the control structure closed (Figure 1(a)). There was no excess water stress in 2009 from other rainfall events so the closed control structure represents the best management practice and mimics setting the tile control structure at 90 cm below the surface level. We set the control structure at 90 cm in 2010, and 85 cm below the surface level in 2011. In 2010 and 2011, with managed control structure levels, there was excess precipitation during the growing season and water was released from the field (Figure 1(b) andFigure 1(c)). Combining the best management practices of retaining water in 2009 while releasing water in both 2010 and 2011 resulted in a significant yield increase (Table 4 and Table 5). This represents

Table 6. Yield, plant height, root disease and leaf disease ratings for cultivars with closed and open tile in 2010 to 2011, Fargo, ND.

		2011	2011	2010	2010 to 2011
Tile	Cultivar	Seed yield	Plant height	Root disease	Leaf disease
		kg ha^{-1}	cm	%	%
Closed	Faller	2198 d[†]	77.8 c	0.7 c	6.9 c
Closed	Howard	2386 bc	78.0 c	5.4 b	4.0 e
Closed	Sabin	2312 c	75.9 c	2.9 bc	4.4 de
Closed	Traverse	2616 a	86.0 a	10.2 a	7.8 b
Open	Faller	2501 abc	81.9 b	1.0 c	6.2 c
Open	Howard	2547 ab	81.3 b	1.8 bc	5.1 d
Open	Sabin	2518 abc	75.7 c	1.0 c	4.9 d
Open	Traverse	2546 ab	86.4 a	3.1 bc	8.7 a
			---Probability > F-----		
Interaction Tile × Cultivar		0.09	0.04	0.07	0.09

[†]Numbers within the same column with at least one similar letter are not significantly different at $P \leq 0.10$.

management level 2 where a control structure would remain closed during dry conditions and open when drainage needed to occur. The results indicate that under dry conditions like 2009 completely open tile may result in lower yields because of the drainage of crop usable water. It also indicates that subsurface water management through a control structure can provide the option to set the water table and conserve plant usable water. To our knowledge this is the first report indicating the importance of managing the water table with a control structure instead of using uncontrolled subsurface drainage in HRSW. Our findings seem to correspond with Nelson et al. [9] , who found that soybean yield for certain varieties was higher with the combination of subsurface drainage and irrigation compared with subsurface drainage alone.

Table 7. Hard red spring wheat yield, plant stand, spikes, height, root disease and Fusarium head blight incidence with closed and open tile, without and with seed treatments in 2009 to 2011, Fargo, ND.

		2009	2010	2011	2009	2009-11	2010-11	2010-11	2010-11	2010-11
					to 2011	Plant		Plant	Root	Fusarium
Tile	Seed treatment	----------------	Seed yield	----------------		stand	Spikes	height	disease	incidence
		----------------	$kg\ ha^{-1}$	----------------		million ha^{-1}	million ha^{-1}	cm	%	%
Closed	No seed treatment	4604 a†	4193 ab	2289 b	3695 a	2.52 a	3.69 a	67.0 a	10.3 a	12.3 b
Closed	Seed treatment	4789 a	4117 b	2468 a	3791 a	2.51 a	3.91 a	69.7 a	6.2 a	12.5 b
Open	No seed treatment	4497 a	4157 ab	2538 a	3731 a	2.61 a	3.61 a	71.8 a	9.0 a	16.4 a
Open	Seed treatment	4511 a	4270 a	2518 a	3766 a	2.53 a	3.65 a	72.4 a	8.4 a	14.2 ab

†Numbers within the same column with at least one similar letter ar not significantly different at $P \leq 0.10$.

3.1.5. Tile × Cultivar

There were only significant interactions between tile treatments and cultivars for yield in 2011, height in 2011, root disease in 2010, and leaf disease in 2010 to 2011, respectively (Table 6). In 2011, the yield of Faller with open tile was significantly higher compared with closed tile. Faller and Howard were significantly taller with open tile compared with closed tile in 2011. In 2010, Traverse had less root disease with open tile compared with closed tile. There were significant differences between cultivars for leaf disease. Across 2010-2011, both Howard and Traverse had more leaf disease with open tile compared with closed tile (Table 6). Traverse had the highest percent leaf disease of the cultivars under open tile. This cultivar was included because of susceptibility to leaf rust and other foliar diseases (Table 2). It may be possible that the taller Howard plants under open tile conditions had a different micro environment resulting in higher leaf disease levels compared with the closed tile condition. However, plant height for Traverse did not differ between closed and open tile (Table 6).

3.1.6. Tile × Seed Treatment

There was no interaction between the tile and seed treatments for yield in the combined analysis (Table 4 and Table 7). The yield with open tile and seed treatment, was 3.7% higher yielding compared with closed tile and seed treatment in 2010. In 2011, open tile without seed treatment was 11% higher yielding than closed tile without seed treatment (Table 7). Cook et al. [30] reported an increase in yield under severe root disease pressure with certain seed treatment combinations that were effective in controlling root disease.

There was no tile by seed treatment interaction and no differences in plant stand, spikes, or plant height (Table 7). The 2009 growing season was dry and FHB and root disease were minimal and therefore, only 2010 to 2011 data are reported. The tile by seed treatment interaction for FHB incidence was significant with higher incidence of FHB when tile was open compared with when it was closed for no seed treatments (Table 7). We expected that open tile would have lower FHB with a lower water table. Possibly there was a difference in the canopy structure and airflow as the HRSW cultivars Faller and Howard were taller on open tile compared with closed tile (Table 6).

3.1.7. Tile × Foliar Fungicide

In 2010, HRSW with foliar fungicide yielded significantly (Table 4) more at 4286 kg ha^{-1} compared with no fungicide application at 4083 kg ha^{-1}. The yield increase may be the result of the significantly lower leaf disease rating with fungicide compared with no application (2.2% vs 3.8%) and

Table 8. Wheat yields with different levels of seed treatments and foliar fungicide treatments in 2009 to 2011, Fargo, ND.

Seed Treatment	Foliar Fungicide Treatment	2010	2011	2009 to 2011
		----------Seed Yield----------		
		--------------kg ha^{-1}---------		
No	No	4124 bc†	2480 ab	3717 b
No	Yes	4226 ab	2347 c	3709 b
Yes	No	4042 c	2463 ab	3702 b
Yes	Yes	4345 a	2523 a	3854 a

†Numbers within the same column with at one least similar letter are not significantly different at $P \leq 0.10$.

significantly lower FHB incidence (8.5% vs 15.8%) and severity (4.5% vs 7.0%) (data not presented). Disease development depends on the environmental conditions, and based on meta-analyses a reduction in disease is expected with a foliar fungicide [31] . However, a fungicide application may only reduce FHB by 50% [31] . There were no significant differences for yield (Table 4) or other observed traits between the tile by foliar fungicide interaction (data not presented).

3.1.8. Seed Treatment × Foliar Fungicide

Although the main objective was to investigate interactions with water management there was an unexpected significant HRSW yield interaction for seed treatment by foliar fungicide in 2010, 2011, and the combined analysis 2009 to 2011 (Table 4). In 2010, the yield with both seed treatment and fungicide was higher than without seed and foliar fungicide treatments (Table 8). In 2011, the yield with seed treatment and fungicide was higher than no seed treatment with foliar fungicide. Overall, across the three years the seed treatment and foliar fungicide combination yielded significantly more than the three other treatments combinations. This seems to indicate that there is a synergism when seed treatment and fungicide application are both used.

4. CONCLUSIONS

With subsurface drainage there can be circumstances, like a dry growing season, when an open tile will not provide the highest yield compared with closed tile. However, controlling the water table (optimum management practice) resulted in significantly higher HRSW yield across years. In 2011, plants of Faller and Howard were taller, and Howard and Traverse had lower root disease severity, and in 2010 to 2011 Howard and Traverse had more leaf disease with open tile compared with closed tile. This seems to indicate that under certain environmental conditions there can be an interaction between cultivars and water management. Across years there was no difference in root disease, stand, number of spikes, plant height and yield response to application of seed treatments with closed or open tile. Only in 2010, there was a 3.7% yield advantage with application of seed treatment with open tile compared with closed tile. Across years there was no positive yield response to application of foliar fungicides, but wheat yield with fungicide was 5% higher than without foliar fungicide in 2010. There was no interaction across years between fungicide application and tile treatments. Across all years, the combination of seed treatment with foliar fungicide resulted in 4% higher yield compared with the other treatments.

In our study, the water table was set at the pre-determined level during the whole season. Future research should investigate crop yield response to actively managing the water table level throughout the season depending on the prevailing weather conditions. In our trial, we applied the fungicide regardless of the current weather outlook. There is a prediction model with probabilities that FHB may occur [1] . In future research, application of inputs should be done based on the likelihood of disease pressure.

ACKNOWLEDGEMENTS

Part of the funding for this research was received from the Minnesota Wheat and Promotion Council and the North Dakota Experiment Station. We gratefully acknowledge the help of Mr. Chad Deplazes, research specialist, and all summer interns assisting in this research.

REFERENCES

1. NDAWN (2014) Weather Data. North Dakota Agricultural Weather Network. North Dakota State University, Fargo. http://www.ndawn.ndsu.nodak.edu/
2. Wiersma J.J., Sands, G.R., Kandel, H.J., Rendahl, A.K., Jin, C.X. and Hansen, B.J. (2010) Responses of Spring Wheat and Soybean to Subsurface Drainage in Northwest Minnesota. Agronomy Journal, 102, 1399-1406.
3. USDA-NASS (2015) http://www.nass.usda.gov/
4. McMullen, M., Jones, R. and Gallenberg, D. (1997) Scab of Wheat and Barley: A Re-Emerging Disease of Devastating Impact. Plant Disease, 81, 1340-1348.
5. Sands, G. (2001) Agricultural Drainage: Soil Water Concepts. University of Minnesota Extension Service. http://www.extension.umn.edu/agriculture/water/agricultural-drainage/soil-water-concepts/
6. Jin, C.X., Sands, G.R., Kandel, H.J., Wiersma, J.J. and Hansen, B.J. (2008) Influence of Subsurface Drainage in Soil Temperature in a Cold Climate. Journal of Irrigation and Drainage Engineering, 134, 83-88.
7. Chieng, S.T., Keng, J. and Driehuyzen, M.G. (1987) Effects of Subsurface Drainage and Subirrigation on the Yields of Four Crops. Canadian Agricultural Engineering, 29, 21-26
8. Kandel, H.J., Brodshaug, J.A., Steele, D.D., Ransom, J.K., DeSutter, T.M. and Sands, G.R. (2013) Subsurface Drainage Effects on Soil Penetration Resistance and Water Table Depth on a Clay Soil in the Red River of the North Valley, USA. Agricultural Engineering International: CIGR Journal, 15, 1-10.
9. Nelson, K.A., Meinhardt, C.G. and Smoot, R.L. (2012) Soybean Cultivar Response to Subsurface Drainage and Subirrigation in Northeast Missouri. Crop Management, 11, 1.
10. Nelson, K.A. and Smoot, R.L. (2012) Corn Hybrid Response to Water Management Practices on Claypan Soil. International Journal of Agronomy, 2012.
11. Nelson, K.A. and Meinhardt, C.G. (2011) Soybean Yield Response to Pyraclostrobin and Drainage Water Management. Agronomy Journal, 103, 1359-1366.

12. Nelson, K.A., Smoot, R.L. and Meinhardt, C.G. (2011) Soybean Response to Drainage and Subirrigation on a Claypan Soil in Northeast Missouri. Agronomy Journal, 103, 1216-1222.

13. McMullen, M. and Adhikari, T. (2009) PP-1249. Fungal Leaf Spot Diseases of Wheat: Tan Spot, Stagonospora nodorum Blotch and Septoria tritici Blotch. NDSU Extension Service, Fargo, 1-6. http://www.ag.ndsu.edu/pubs/plantsci/pests/pp1249.pdf

14. Bechtel, D.B., Kaleikau, L.A., Gaines, R.L. and Seitz, L.M. (1985) The Effects of Fusarium graminearum Infection on Wheat Kernels. Cereal Chemistry, 62, 191-197.

15. Ransom, J.K. and McMullen, M.P. (2008) Yield and Disease Control on Hard Winter Wheat Cultivars with Foliar Fungicides. Agronomy Journal, 100, 1130-1137.

16. Harding, H. (1978) Root Rot of Cereals-Everyone's Problem? Canada Agriculture, 23, 25-29.

17. Machacek, J.E. (1943) An Estimate of Loss in Manitoba from Common Root-Rot in Wheat. Scientia Agricola, 24, 70-77.

18. Stack, R.W. and McMullen M.P. (1991) Effects of Fungicide Seed Treatments on Common Root Rot of Spring Wheat and Barley. North Dakota Farm Research, 49, 13-16.

19. Wiersma, J.J. and Kandel, H.J. (2004) The Response of Fusarium graminearum-Infected Seed of Hard Red Spring Wheat to Vitavax Extra RTU and Dividend XL Seed Treatments. Plant Health Progress.

20. Zadoks, J.C., Chang, T.T. and Konzak, C.F. (1974) A Decimal Code for the Growth Stages of Cereals. Weed Research, 14, 415-421.

21. USDA-NRCS (2012) Soil Survey of Cass County. USDA-NRCS, Washington DC.

22. Anderson, J., Wiersma, J., Linkert, G., Reynolds, S. and Springer, C. (2010) Varietal Trial Results: Wheat, Hard Red Spring. Minnesota Agricultural Experiment Station, University of Minnesota, Minneapolis and Saint Paul. http://www.maes.umn.edu/sites/maes.umn.edu/files/wheat_spring_2009.pdf

23. Ransom, J.K., Mergoum, M. and Simsek, S. (2009) A-574. North Dakota Hard Red Spring Wheat Variety Trial Results for 2009. NDSU Extension Service, Fargo.

24. Wiersma, J.J. and Ransom, J.K. (2005) The Small Grains Field Guide. North Dakota State University Extension Publication A290 and University of Minnesota Extension Service Publication 0788-S, Fargo and St. Paul.

25. Nowatzki, J., Halley, S., Van Ee, G., Hofman, V., McMullen, M., Hollingsworth, C. and Ruden, B. (2013) AE1314. Ground Application of Fungicide for the Suppression of Fusarium Head Blight in Small Grains. NDSU Extension Service, Fargo, 1-4. http://www.ag.ndsu.edu/pubs/ageng/machine/ae1314.pdf

26. Stack, R.W. and McMullen, M. (1999) PP-785. Root and Crown Rots of Smalls Grains. North Dakota State University, Fargo. http://www.ag.ndsu.edu/pubs/plantsci/smgrains/pp785w.htm

27. Tobias, D.J., Stack, R.W., Puri, K.D., Riveland, N. and Zhong, S. (2009) Reactions of Hard Red Spring Wheat to Common Root Rot under Field Conditions of Northern United States of America. Euphytica, 167, 165-172.

28. Stack, R.W. and McMullen, M.P. (2011) PP-1095. A Visual Scale to Estimate Severity of Fusarium Head Blight in Wheat. North Dakota State University Extension Publication, Fargo. http://www.ag.ndsu.edu/pubs/plantsci/smgrains/pp1095.pdf

29. Poole, C.A., Skaggs, R.W., Cheschier, G.M., Youssef, M.A. and Crozier, C.R. (2013) Effects of Drainage Water Management on Crop Yields in North Carolina. Journal of Soil and Water Conservation, 68, 429-437.

30. Cook, R.J., Weller, D.M., El-Banna, A.Y., Vakoch, D. and Zhang, H. (2002) Yield Response of Direct-Seeded Wheat to Rhizobacteria and Fungicide Seed Treatments. Plant Disease, 86, 780-784.

31. Willyerd, K.T., Li, C., Madden, L.V., Bradley, C.A., Bergstrom, G.C., Sweets, L.E., McMullen, M., Ransom, J.K., Grybauskas, A., Osborne, L., Wegulo, S.N., Hershman, D.E., Wise, K., Bockus, W.W., Groth, D., Dill-Macky, R., Milus, E., Esker, P.D., Waxman, K.D., Adee, E.A., Ebelhar, S.E., Young, B.G. and Paul, P.A. (2012) Efficacy and Stability of Integrating Fungicide and Cultivar Resistance to Manage Fusarium Head Blight and Deoxynivalenol in Wheat. Plant Disease, 96, 957-967.

Index